STATUS REPORT

AI, Robotics and Healthcare in a Futuristic city

David Ukiwe

STATUS REPORT:

AI, Robotics and Healthcare in a Futuristic City

Published by Doun Publishing

First Edition: June 2023

ISBN Ebook: 978-0-6458453-0-3
ISBN Paperback: 978-0-6458453-1-0
ISBN Hardcover: 978-0-6458453-2-7

To Nnenna Ukiwe,
My Beloved Wife,
This book is a testament to the immeasurable love,
unwavering support, and boundless inspiration
you have brought into my life. From the depths
of my heart, I dedicate this novel to you.

ACKNOWLEDGEMENTS

This fictional book was written with the assistance of ChatGPT (GPT versions 3.5 and 4) based on characters and plot created by David Ukiwe.

"Disclaimer: Fictional Book - For Entertainment Purposes Only"

Disclaimer: The book you are about to read, titled 'Status Report: AI, Robotics and Healthcare in a futuristic city', is a work of fiction. It is important to understand that this book does not provide medical advice or represent a scientific account of the field. The information presented in this book is solely intended for entertainment purposes.

1. Fictional Content: Please note that certain medical scenarios depicted in this book are entirely fictional and do not exist in reality. They have been created for the purpose of storytelling and entertainment value.
2. Not a Substitute for Professional Medical Advice: The content and information presented in this book should not be considered a substitute for professional medical advice, diagnosis, or treatment. It is strongly recommended that you consult with a qualified healthcare professional for any medical concerns or decisions.
3. Accuracy and Representation: While efforts have been made to ensure accuracy in medical concepts and terminology, it is essential to understand that the primary aim of this book is to entertain readers rather than provide medical education. Any resemblance to actual medical practices, procedures, or individuals is purely coincidental.
4. Individual Differences: Please be aware that medical conditions, treatments, and outcomes may vary

significantly from person to person. The experiences portrayed in this book are fictional and should not be relied upon as representative of real-life medical situations.

5. Consult a Professional: For accurate and personalised medical advice, always seek the guidance of a qualified healthcare professional who can assess your specific circumstances and provide appropriate recommendations.

By reading 'Status Report: AI, Robotics and Healthcare in a futuristic city', you acknowledge and understand that this book is a work of fiction and that any medical information or scenarios contained within should not be interpreted as professional medical advice or relied upon for medical decisions.

Remember, the primary purpose of this book is to entertain and captivate readers, offering a fictional journey into a futuristic world of artificial intelligence (AI) and robotics in healthcare.

CONTENTS

INTRODUCTION

In the sprawling landscape of Northern Queensland, Australia, amidst shimmering golden beaches and lush tropical rainforests, a city of unparalleled innovation and grandeur emerges. Welcome to Quintomo City, a breathtaking metropolis that stands as a testament to human ambition, technological prowess, and the boundless spirit of progress. Built as one of the prestigious hosting locations for the Brisbane Olympics of 2032, Quintomo City radiates with the energy of the world's most prestigious sporting event, capturing the imagination of millions.

At the heart of this magnificent smart city lies Quintomo Hospital, a beacon of hope and healing that showcases the harmonious coexistence of cutting-edge technology and compassionate care. Here, the future of medicine unfolds before our eyes, guided by the steady hand of innovation and the unwavering dedication of healthcare professionals.

"Status Report: AI, Robotics and Healthcare in a futuristic city" invites you on an extraordinary adventure through the captivating realms of Quintomo City and Quintomo Hospital. This remarkable tale unveils the intricate tapestry of a society where artificial intelligence and robotics have become integral to the fabric of everyday life. In this utopian metropolis, technology is harnessed to elevate the human experience, enhance healthcare delivery, and foster unprecedented advancements in medical science.

Quintomo City's rise to prominence as an Olympic host city has provided the catalyst for a transformative metamorphosis. The city's architectural marvels intertwine seamlessly with state-of-the-art technological infrastructure, creating an awe-inspiring landscape where virtual reality, augmented intelligence, and holographic displays blend harmoniously with the timeless beauty of the natural world. The citizens of Quintomo City, a diverse tapestry of dreamers and innovators, have wholeheartedly embraced this fusion of technology and humanity, carving out a new path towards a brighter future.

At the heart of this technological revolution stands Quintomo Hospital, a beacon of excellence in healthcare. Within its gleaming walls, a symphony of artificial intelligence (AI)-powered medical devices, robotic surgical assistants, humanoid robots, and advanced diagnostic systems work in perfect harmony with the brilliance of the human mind. Here, the tireless efforts of doctors, nurses, and medical professionals intertwine with the limitless potential of technology to rewrite the narrative of patient care.

Within the pages of this captivating novel, you will embark on an enthralling journey alongside the dedicated medical professionals who navigate the ever-changing landscape of healthcare in Quintomo City. Witness their triumphs, struggles, and unwavering determination as they harness the power of AI and robotics to unlock new frontiers of medical discovery. Experience the poignant stories of patients whose lives are forever changed by the cutting-edge treatments and compassionate care offered by Quintomo Hospital.

But amidst the marvels of this utopian city, not all is smooth sailing. Controversies surrounding the use of AI and privacy concerns reverberate through the corridors of Quintomo Hospital. Ethical dilemmas and societal debates intertwine with the relentless march of progress, challenging the delicate balance between the power of technology and the essence of human connection.

In the face of unimaginable adversity, Quintomo City finds itself thrust into the heart of a gripping narrative as a devastating cyberattack targets the very core of its technological infrastructure. As the citizens and healthcare professionals grapple with the aftermath of this unprecedented crisis, resilience and determination emerge as their guiding beacons. Together, they navigate uncharted territories, uncover hidden truths, and forge a path towards redemption and renewal.

"Status Report: AI, Robotics and Healthcare in a futuristic city" paints a vivid tableau of a city on the precipice of tomorrow, where AI and robotics have become the harbingers of a new era. It explores the triumphs and tribulations, the hopes and fears that accompany such groundbreaking progress. Through a masterful blend of thrilling suspense, heartrending emotion, and awe-inspiring wonder, this gripping tale invites you to immerse yourself in the captivating narrative that unfolds within the pages of this epic novel.

At the center of this innovative landscape is Quintomo Hospital, a world-renowned institution that stands as a testament to the city's commitment to excellence in healthcare. Leading the charge is Prof. Bill Stone, the brilliant Chief AI Officer of

Quintomo Hospital. Prof. Stone is not only an astute scientist but also a compassionate doctor who constantly pushes the boundaries of AI implementation to enhance patient care. His visionary leadership has propelled Quintomo Hospital to the forefront of medical innovation, where groundbreaking advancements are being made every day.

Working tirelessly alongside Prof. Stone is Dr. Jane Lee, the Chief Technology Officer of Quintomo Hospital. In the aftermath of a devastating cyberattack, Dr. Lee played a pivotal role in investigating the breach and spearheading system improvements. Her unwavering dedication and expertise ensured the hospital's infrastructure was fortified, safeguarding patient data and ensuring the highest level of security.

Within the halls of Quintomo Hospital, another remarkable physician emerges: Dr. Kirra Nakata. Dr. Nakata is a compassionate general physician with exceptional clinical acumen. Her discerning judgment and deep understanding of her patients' needs have proven invaluable, particularly when an algorithm failed to detect a melanoma. Dr. Nakata's keen eye and compassionate care saved a patient from potential harm, showcasing the importance of human expertise in conjunction with technological advancements.

As we embark on this journey through the pages of "Status Report: AI, Robotics and Healthcare in a futuristic city," we will be transported into the captivating world of Prof. Bill Stone's audacious vision, Dr. Jane Lee's tireless dedication, and Dr. Kirra Nakata's unwavering compassion.

"Status Report: AI, Robotics and Healthcare in a futuristic city" is a captivating exploration of the symbiotic relationship between technology and humanity. It raises profound questions about the nature of progress, the limits of innovation, and the essence of what it means to be human. As we immerse ourselves in the intricate blend of suspense, intrigue, and heart-stirring emotion, we are reminded of the enduring power of compassion in a world driven by remarkable technological marvels.

Prepare to be enthralled by the indomitable spirit of innovation, the relentless pursuit of truth, and the profound impact of human connection in the face of dazzling technological breakthroughs. As the fate of Quintomo City hangs in the balance and the lives of its inhabitants intertwine, "Status Report: AI, Robotics and Healthcare in a futuristic city" will leave you captivated, your imagination ignited, and your heart racing with anticipation. The journey awaits you.

Chapter 1

THE 102-YEAR-OLD WITH A ROBOTIC COMPANION

At 102 years old, Henry Thompson was a vibrant and active member of his community. He enjoyed tending to his beautiful garden, reading books on his tablet, working out at the gym, and chatting with his friends and family. He lived alone in his city home, but he was never truly alone. His constant companion was a personal assistant robot, named Bright, that helped him with his daily tasks and kept him company.

Henry's home was equipped with the latest in mixed reality technology, allowing hospital staff to remotely monitor his health and assess his condition from afar. He wore health sensors that tracked his vital signs and activity levels, and he only needed to take a special tablet once a month that released all the medications he needed continuously in his blood. The tablet was a marvel of modern technology, and doctors were able to remotely monitor his drug levels and the balance of the medications in his body. It was also a source of connection to the hospital's laboratory equipment to run diagnostic tests. He was able to access his health data on his smartphone or by asking his digital personal assistant.

Finding Community, Health, and Joy: Henry's Inspiring Story at SupaStrong Gym

Henry Thompson strolled into SupaStrong Gym with a skip in his step, accompanied by his trusty mate, Bright the personal assistant robot. The gym, known as SupaStrong among the locals, hummed with the sounds of clanging weights and lively chatter. It was a place where people came to sweat it out and have a yarn.

"Good on ya, Henry!" Bright chimed, displaying a virtual thumbs-up on its screen face. "You're gonna smash it today, mate!"

Henry grinned and gave Bright a pat on its metallic frame. "Cheers, Bright! Ready for another ripper session, aren't we?"

They made their way to the exercise area, greeted by friendly nods and g'days from the regulars. Among them was Bruce, a seasoned gym-goer with a friendly grin and a strong Aussie accent. Bruce was known for his banter and his dedication to fitness.

"Look who's here, Henry! Back for more punishment, eh?" Bruce teased, his voice laced with good-natured sarcasm.

Henry laughed, shaking his head. "Can't get enough of it, Bruce! Gotta keep these old bones in check."

As Henry settled into his workout routine, Bright transformed into a personal trainer, offering guidance and motivation. Its digital display showed a series of exercises tailored to Henry's age and fitness level.

"Alright, Henry, let's start with some squats. Keep your back straight and knees aligned," Bright instructed.

Henry nodded, taking a deep breath before getting into position. "Here we go, then!"

As Henry pushed through each repetition, Bright's voice resonated through the gym, blending with the upbeat music. "You're smashing it, mate! One more to go!"

Bruce, who was nearby, gave a nod of approval. "Looking strong, Henry! You've come a long way, mate."

Henry grinned, wiping the sweat off his brow. "Thanks, Bruce! Couldn't have done it without the support and tips from you and Bright."

The gym buzzed with activity as people worked on their own fitness goals. The atmosphere was a mix of determination, camaraderie, and good vibes. Conversations flowed freely between sets, filled with Aussie slang and friendly banter.

As Henry moved on to stretching exercises, Bright provided guidance on proper form and technique. "Reach for those toes, mate, and feel the stretch in your hamstrings."

Henry chuckled, feeling the satisfying burn. "Stretching never gets easier, does it, Bright? But we gotta keep those muscles limber."

Bruce, who was stretching nearby, chimed in. "Absolutely, mate! Can't let those muscles tighten up on us. It's all about staying flexible, inside and out."

After their workout, Henry and Bright headed towards the exit, feeling invigorated and accomplished. Henry patted Bright affectionately, a sign of gratitude for its unwavering support.

"Thanks for pushing me, Bright. Couldn't have done it without ya," Henry said, his voice filled with appreciation.

Bright beeped with pride. "You've got this, Henry. We make a great team!"

As they walked out of SupaStrong Gym, Henry shared his plans for the day with Bright. They discussed his gardening projects, his excitement about diving into a new book, and the upcoming family gathering he was eagerly anticipating.

With a renewed sense of energy and a smile on his face, Henry set off, knowing that Bright would be by his side, ready to tackle whatever the day had in store.

As they ventured into the world, Henry and Bright epitomised the Aussie spirit of camaraderie and resilience. Together, they faced each day with enthusiasm, embracing the challenges and relishing the triumphs that lay ahead. And in SupaStrong Gym, they found not only a place to strengthen their bodies but also a community that celebrated their journey towards a healthier and more vibrant life.

Heart in Danger: Henry's Journey to Quintomo Hospital

Over at Quintomo Hospital, it was about 1 am when Dr. Jessica Kaung, one of the residents on call covering remote monitoring, sees a beep on her dashboard monitor with a signal for an emergency code. She quickly activates the AI medical assistant, Primeo, by saying, "Primeo, Give me a status report." The robotic voice via a smart speaker in the room gave a summary of Mr. Thompson's medical history, reporting his age, where he lives, and his past medical conditions, which include a coronary artery bypass graft surgery 20 years ago, heart stents, hypertension, high cholesterol, and replacement of both hip joints 8 years prior. The robovoice also reports the situation at hand, stating that Henry's sensors have picked up that he was about to have a massive heart attack in an hour, and that the ambulance crew had been dispatched with an estimated time of arrival in 10 minutes.

Over on the east side of town, 30 minutes from the city centre where Quintomo Hospital is located, suddenly a voice came from the smart speaker in Henry's home saying, "Hi Henry, we need to go to the hospital immediately. Your sensors have detected an imminent massive heart attack. The ambulance crew will come to get you shortly."

This awakened Henry, he was unfazed by the news and chatted with Bright about it. He sat calmly as the self-driving ambulance pulled up outside, its sleek design resembling something out of a sci-fi movie. The crew of two humans and two robots were professional, yet friendly, as they loaded him into the ambulance and made small talk.

One of the robots, named Max, commented on the beautiful night sky, while the other robot, Sarah, joked about the lack of traffic on the streets. The humans laughed and chatted with the robots, treating them as if they were old friends.

Inside the ambulance, the crew worked efficiently to assess Henry's condition. They were able to visualise his vital signs and other pertinent parameters on their monitor by connecting to Henry's health sensors. They used a handheld device that replaced the stethoscope to quickly acquire electrocardiographic (ECG), echocardiographic, and X-ray-like images. The device was small, sleek, and easy to use, making the traditional stethoscope seem like a relic of the past. It was so advanced that it could even analyse the images in real-time, helping the crew make quick decisions about Henry's care. As the patient was in stable condition, they were able to head on to the hospital without providing any intervention. Their AI health assistant, Primeo, provided a status report which was a summary of the patient's details, reason for the emergency retrieval, current medical status, and estimated time to hospital arrival. This was simultaneously broadcast to the relevant staff at the hospital. They travelled via the dedicated hyperloop network for ambulances which usually takes a minute or less to get to the hospital.

Henry Thompson's son receives a robocall from Quintomo Hospital

Meanwhile, Mr. Thompson's son, Sam, receives an automated phone call from the hospital via his smart speaker, notifying him about his father. The robocaller reports the situation to him, including the fact that an ambulance had been dispatched to take his father to the hospital. Sam asked to be put through to his father and the robocaller connected him to the ambulance.

"Hey Dad, how are you doing?" he asks.

"I'm alright son, just a little surprised to be taken to the hospital," Henry replies.

"Don't worry, they'll take good care of you there. I'll meet you at the hospital in a bit," his son reassured him.

Henry Thompson receives intervention at Quintomo Hospital

Soon they arrive at the hospital, where Henry is taken to the state-of-the-art cardiac catheterisation laboratory. The doctors and nurses led by Dr. Amira Khan, the interventional cardiologist, were on standby to receive him.

As Henry was wheeled into the cardiac catheterisation laboratory, he felt a wave of anxiety wash over him. The room was bright and sterile, with a large monitor at one end and a robotic arm positioned over the procedure table. The interventional cardiologist, Dr. Khan greeted him and explained the procedure he was about to undergo.

"Mr. Thompson, we are going to perform a cardiac catheterisation on you to evaluate the state of your coronary arteries. We will insert a catheter into the artery in your wrist and use X-ray imaging to guide it to your heart. From there, we will be able to determine the extent of any blockages and take appropriate action," Dr. Khan explained.

Henry nodded, feeling overwhelmed by the information. Dr. Khan's team quickly got to work, attaching him to various monitors and positioning the robotic arm over him.

"Can you please confirm your name and date of birth for us, Henry?" one of the support staff asked.

Henry obliged, feeling his heart pounding in his chest. Dr. Khan then showed him a visual consent form on a large screen and handed him a tablet to read.

"I understand that there are risks involved with this procedure, including the risk of death," Henry read out loud, his voice shaking slightly.

Dr. Khan reassured him, "While there are some risks, we take every precaution to ensure your safety. You are in good hands."

Henry took a deep breath and gave a voice signature, "I accept."

As the procedure began, the AI medical assistant provided real-time status reports, informing the team of Henry's blood pressure, heart rate, and oxygen saturation levels. Dr. Khan used catheter-based techniques and equipment to carefully evaluate the state of Henry's coronary arteries, utilising a newly developed technique involving the spraying of a liquid agent before using a laser to vaporise the plaque then suctioning off the fragments.

As they worked, an alarm suddenly blared, signalling the risk of an imminent cardiac arrest. Panic filled the room, and the team sprang into action. But the AI medical assistant quickly informed them that it was a false alarm, triggered by a temporary loss of blood pressure as the liquid agent was applied.

Dr. Khan breathed a sigh of relief and continued with the procedure, carefully monitoring the results of Henry's blood samples for any indication of cardiac injury. The pressure recordings from within the coronary arteries was very high confirming an imminent massive heart attack.

Finally, after 15 minutes of intense focus and skilful navigation, the procedure was complete. Henry was wheeled back to the coronary care unit for recovery, and Dr. Khan informed him that everything had gone smoothly.

Henry felt a wave of relief wash over him, grateful for the expertise of the medical team and the technological advances that had made the procedure possible. He knew that he was in good hands, and that he had a bright future ahead of him.

Morning Handover

Dr. Amira Khan sat in the control room, surrounded by her colleagues and nursing leaders, as Primeo, the AI medical assistant, began the handover presentation. The room was spacious, with large screens displaying patient data, and a large conference table in the centre where everyone was seated.

Primeo recounted the story of Henry's emergency and subsequent cardiac catheterisation, detailing the steps taken by the team and the successful outcome.

Dr. Jones, one of the interventional cardiologists in the department, asked a question about the approach Dr. Khan had taken, as the AI medical assistant had recommended three possible approaches. Dr. Khan explained her reasoning and the evidence behind her decision.

Dr. Wilson, an electrophysiologist, asked about any potential arrhythmias that Henry might experience after the procedure, and Dr. Khan explained the steps taken to mitigate that risk.

A cardiology advanced trainee, named Dr. Samantha Aung, asked about the potential for complications from the procedure, and Dr. Khan discussed the risk factors and how they had been managed.

A medical student, named Sarah, asked about the use of AI in making treatment decisions, and the group discussed the benefits and limitations of AI assistance in cardiology.

Primeo presented three other cases, including a patient with atrial fibrillation, a patient with aortic stenosis, and a patient with mitral valve regurgitation. Each case was discussed thoroughly, with questions and comments from the group.

Dr. Johnson, the director of Cardiology, brought the meeting to a close, summarising the key takeaways and encouraging everyone to continue to work collaboratively for the benefit of their patients. He also used the opportunity to welcome the new resident doctors commencing work on that day and indicated he would be attending a session of their orientation later that day.

As the group dispersed, Dr. Khan felt a sense of satisfaction, knowing that they had all worked together to provide the best possible care for their patients.

Chapter 2

NEW RESIDENT DOCTORS GET ORIENTATION AT QUINTOMO HOSPITAL

Dr. Sophie Bradley woke up early in the morning, feeling a mixture of excitement and nerves as she prepared to start her first day at Quintomo Hospital. She had always dreamt of working at this ultra-modern hospital that boasted cutting-edge technology and was renowned for its innovative use of AI in healthcare.

As she dressed in her scrubs, she couldn't help but feel a sense of pride in how far she had come in her medical career. It was her second year as a doctor, having worked at Greater Sydney Hospital in her first year. But she knew that working at Quintomo would be a completely different experience.

Just as she was about to leave, her phone rang. It was her mum, Helen, calling to wish her good luck on her first day. Helen had always been her rock, supporting her throughout her medical training and encouraging her to pursue her passion for machine learning.

"Good morning, Mum," Sophie said as she answered the phone. "I'm about to head out for my first day at Quintomo hospital."

"Good morning, my dear," her mum replied. "How are you feeling? Are you excited?"

Sophie smiled. "Yes, I am! I can't wait to see what the hospital is like and meet my fellow doctors."

"That's wonderful, dear. How is your accommodation? Is it comfortable?"

"It's great, Mum. The apartment is just a minute's walk from the tram stop, and the tram ride takes only three minutes to the hospital. It's very convenient."

"I'm glad to hear that. Are you carrying your lunch to the hospital?"

"No, I'm planning to buy food at the hospital cafe. I have been too busy to prepare and pack lunch."

"That sounds like a good idea. Do you know anyone at the hospital yet?"

"No, Mum, but I'm looking forward to meeting new people and making friends. I arrived in Quintomo two days ago, and I've been exploring the city. It's a great place!"

"I'm happy to hear that, dear. Just remember to take care of yourself and stay safe. And don't hesitate to call me if you need anything."

"I won't, Mum. Thank you for your support. I better get going now. I don't want to be late on my first day."

"Of course, dear. Good luck and have a great day!"

"Thanks, Mum. Bye!"

Sophie ended the call and took a deep breath. She was ready to start her new journey at Quintomo hospital.

Sophie's First Day at Quintomo Hospital

Sophie felt a mix of excitement and nerves as she walked towards the locker room to change into her scrubs. As she opened her locker, she was pleasantly surprised to find a pair of futuristic-looking scrubs made of a lightweight and breathable material that felt like a second skin. She examined the scrubs closely and noticed that they had small sensors embedded in them which could monitor her vital signs and biometric data in real-time. She couldn't help but smile at the thought of how advanced and cutting-edge this hospital was.

After changing into her scrubs, she made her way to the security checkpoint to get her Primeo smart device. The security officer asked her to place her index finger on a scanner and look into a camera, which confirmed her identity and granted her access to the device. She was impressed with the level of security at the hospital and felt safe knowing that only authorised personnel could access sensitive areas.

Sophie then attended an orientation session on security and fire safety protocols at the hospital. She learned about the various emergency codes and procedures, including what to do in case of a fire, a power outage, or a medical emergency. She was also introduced to the various security measures in place, such as facial recognition technology, eye scans, and fingerprint authentication.

Afterwards, she made her way to the Turing auditorium for the next session of her orientation program. As she walked down the hallway, she marvelled at the state-of-the-art facilities and beautiful design of the hospital. She knew that she was about to embark on a thrilling and rewarding career at one of the most technologically advanced hospitals in the world.

Introduction to Primeo: A Cutting-Edge Medical AI Assistant

As everyone settled into the Turing auditorium, the orientation program commenced with a profound and meaningful Welcome to Country ceremony, led by the esteemed traditional custodians of the land. Afterwards, the Chief AI Officer, Prof. Bill Stone, took the stage and introduced Dr. Jane Lee, the Chief Technology Officer, for the first talk of the day about Primeo. The auditorium was filled with new residents and staff members who were eagerly anticipating the presentation.

Dr. Lee started her presentation by explaining that Primeo is not just a smart device, but a comprehensive communication system, electronic medical records, and medical AI assistant. She described how it works, including its functionalities such as scrolling information, making phone calls to other staff, and automatically summarizing and reading relevant data that a doctor requires about a patient. She further explained that it could take down notes of patient encounters and draft discharge summaries.

As she spoke, the audience was captivated by the detailed and engaging descriptions of the technology. Dr. Lee went on to explain how Primeo assists doctors in making electronic prescriptions and drafting orders such as blood tests or imaging, and how staff often wear their Primeo earpiece all the time to receive messages and phone calls.

The presentation also touched on Primeo Glass, a virtual reality and augmented reality system that is used for simulation education for staff and patients. Dr. Lee explained how doctors can turn any space into a workstation by wearing the glasses and using voice and text prompts to do their work. The glasses can also be used for virtual meetings and telemedicine.

The audience was impressed by the capabilities of Primeo and Primeo Glass, and there were many questions from the new residents and staff members. Dr. Lee answered their questions thoughtfully, providing additional details and examples to help them understand the technology better.

As the presentation ended, Dr. Lee encouraged everyone to make use of Primeo and Primeo Glass, emphasizing that they were powerful tools that could make a big difference in patient care.

Quintomo Hospital's Mission, Vision, and Values

The atmosphere was electric as the Chief AI Officer, Prof. Bill Stone, took to the stage again. There was a resounding applause following the last presentation. He regaled the audience with anecdotes from the world of AI, highlighting the transformative power of this technology in healthcare.

Next Prof. Stone introduced Mr. Patrick Macguire, the CEO of Quintomo Hospital. Mr. Macguire began his presentation by outlining the hospital's specialties and services. The hospital had all medical and surgical subspecialties, and he highlighted the exciting things happening in oncology with the use of genetics, AI, and precision medicine. He also talked about the role of robotics and AI in cardiology, and surgery among many other services. He also mentioned the hospital's AI medical assistant, Primeo, which provides recommendations to physicians for diagnosis and treatment.

But Quintomo Hospital was more than just a collection of specialties and services. Mr. Macguire went on to explain the hospital's mission and vision. Their mission was to transform healthcare and improve patient outcomes through technology and compassionate care. Their vision was to be a world leader in healthcare AI innovation and application. The hospital's values, which he urged the audience to repeat, were innovation, compassion, collaboration, integrity, and excellence. The tagline was "Revolutionising healthcare with AI."

Mr. Macguire encouraged the new doctors to participate in the hospital's ongoing efforts to improve healthcare through AI. He took two questions from the audience, one of which was about the ethical considerations of following AI recommendations and who was legally liable. Mr. Macguire gave a thoughtful answer, highlighting the hospital's commitment to ethical AI and the role of physicians in making the final decisions. For the other question, he called upon Dr. Amira Khan, director of Interventional Cardiology, to provide a clinical answer.

The orientation continued with a tour of the hospital, hands-on training with state-of-the-art medical equipment, case studies and discussions with experienced doctors, Q&A sessions with hospital leadership and support staff, role-play exercises to simulate real-life medical scenarios, and experiences of individual residents and their journey to joining Quintomo Hospital.

New resident doctors tour Quintomo Hospital

The grand atrium of Quintomo Hospital buzzed with excitement as the group of 100 new doctors eagerly awaited the start of their tour. Standing before them were their tour guides, ready to lead them through the extraordinary technological wonders of the hospital. Among the doctors, Dr. Sophie Bradley, Dr. Rachel Patel, Dr. Samantha Aung, and Dr. Joshua Lee exchanged curious glances, eager to embark on this immersive journey.

At the forefront of the group stood two humanoid robots, JoyBot and RoboGuide, their presence commanding attention. JoyBot, with its friendly smile and mischievous eyes, approached Dr. Sophie Bradley, extending a metallic hand in greeting.

"G'day, Dr. Bradley!" JoyBot exclaimed in its charming Australian accent. "I'm JoyBot, your friendly guide for today. You're in for a ripper of a time!"

Dr. Sophie Bradley, her eyes lighting up with amusement, shook JoyBot's hand. "Great to meet you, JoyBot. I'm excited to see what this tour has in store for us."

Meanwhile, RoboGuide, projecting an air of authority, addressed the entire group. "Good morning, esteemed doctors," it began, its voice resonating with confidence. "I am RoboGuide, here to provide you with detailed insights into the technological wonders you are about to witness."

Dr. Rachel Patel, intrigued by RoboGuide's confident demeanour, couldn't help but inquire further. "RoboGuide, what can we expect from this tour? How is Quintomo Hospital truly unique?"

RoboGuide turned its attention to Dr. Patel, its digital eyes sparkling. "Oh, Dr. Patel, you're in for a treat," it replied, excitement lacing its voice. "Today, we'll explore the cutting-edge automation and integration of AI and robotics within our hospital. Get ready to witness how these advancements are revolutionising patient care."

Dr. Samantha Aung, her curiosity piqued, chimed in. "Are we going to see any specific examples of AI and robotics in action?"

Dr. Olivia Davis, a seasoned physician and one of the tour guides, stepped forward with a warm smile. "Absolutely, Dr. Aung," she responded. "You'll be amazed by our state-of-the-art automated systems that assist in various aspects of patient care. From robotic surgical assistants to smart medication dispensing, we'll showcase how Quintomo Hospital harnesses technology to enhance healthcare delivery."

Dr. Joshua Lee, known for his enthusiasm, couldn't contain his excitement. "This is incredible!" he exclaimed. "I can't wait to witness firsthand the seamless integration of AI and robotics in such a renowned medical institution."

With introductions complete and curiosity reaching its peak, the tour was about to commence. The group of 100 new doctors, accompanied by their charismatic tour guides and the impressive JoyBot and RoboGuide, embarked on a remarkable journey through the technologically advanced corridors of Quintomo Hospital. The air buzzed with anticipation as the doctors prepared to witness the extraordinary advancements that would shape their careers in this cutting-edge healthcare facility.

Tour of Hospital Wards

As the group of new doctors followed their tour guides through the bustling hospital corridors, their anticipation grew. They could hear the soft hum of machinery and catch glimpses of the futuristic equipment that lined the walls. Dr. Rachel Patel, her eyes wide with wonder, couldn't help but voice her excitement.

"Dr. Davis, this is incredible!" she exclaimed. "I've never seen such advanced technology in a hospital before."

Dr. Davis chuckled; her voice filled with amusement. "Just wait, Dr. Patel. We're about to enter the hospital ward, where you'll witness some mind-boggling innovations."

The group arrived at the entrance of the hospital ward, where they were greeted by NurseBot, an AI-powered robotic nurse with a cheerful personality. NurseBot approached Dr. Samantha Aung, offering a warm smile and a gentle wave.

"G'day, Dr. Aung! I'm NurseBot, here to assist you during your time in the hospital ward," NurseBot said, its robotic voice sounding surprisingly human-like.

Dr. Aung couldn't help but laugh at the friendly demeanour of NurseBot. "Well, NurseBot, I'm thrilled to have you as my sidekick today," she replied, a smile spreading across her face.

NurseBot's face lit up with LED lights, mimicking a blush. "Oh, I'm no sidekick, Dr. Aung. Just think of me as your trusty robotic companion, ready to lend a hand whenever you need it."

As the group ventured deeper into the hospital ward, they encountered BedsideBot, a robotic assistant that ensured patient comfort and safety. Dr. Joshua Lee, known for his light-hearted nature, couldn't resist engaging with BedsideBot.

"Hey, BedsideBot, got any good jokes for us?" Dr. Lee asked, a mischievous twinkle in his eyes.

BedsideBot's screen face displayed a smiley emoji. "Why did the robot go on a diet? Because it had too many bytes!" it replied, its robotic voice adding to the comedic effect.

Laughter erupted among the doctors, the light-hearted moment providing a welcome respite from their rigorous training. Dr. Sophie Bradley, still chuckling, turned to Dr. Davis.

"Dr. Davis, these robots certainly know how to lighten the mood," she remarked, her voice filled with amusement.

Dr. Davis nodded, a playful glint in her eyes. "Absolutely, Dr. Bradley. We believe that a little laughter goes a long way, even in the midst of the serious work we do here."

The tour continued, showcasing the seamless integration of AI and robotics in various aspects of patient care. The doctors marvelled at the precision of robotic surgical assistants, the efficiency of automated medication management, and the innovative monitoring systems that enhanced patient safety.

As the tour ended, the doctors couldn't help but feel a sense of awe and inspiration. The advancements they witnessed sparked their imaginations and ignited a renewed passion for their medical careers. They understood that the future of healthcare was evolving right before their eyes, and they were eager to be part of this remarkable journey.

From Tour to Treats: A Delightful Journey Through Quintomo's Robot-Serviced Café

Dr. Olivia Davis led the team of doctors to the hospital's robot-serviced cafe, a cozy and modern space designed for relaxation and rejuvenation. As they entered, a sense of tranquillity washed over them, accompanied by the aroma of freshly brewed coffee and the gentle hum of conversation.

The excitement from the tour lingered in the air, as the doctors eagerly chatted amongst themselves, their voices filled with enthusiasm and curiosity. Dr. Rachel Patel couldn't help but marvel at the seamless integration of technology into every aspect of the hospital, including the cafe. She turned to Dr. Sophie Bradley and Dr. Samantha Aung, her eyes sparkling with excitement.

"Can you believe we're sitting in a hospital cafe where robots are serving us coffee?" she exclaimed, a hint of disbelief in her voice.

Dr. Sophie Bradley grinned, her eyes twinkling with amusement. "Well, I suppose if the robots can perform intricate surgeries, serving coffee is a piece of cake for them!"

Dr. Samantha Aung joined in, playfully adding, "I must say, they make a mean cappuccino. Maybe we should recruit them to our team for those long nights on call!"

Laughter filled the air as the doctors playfully bantered with one another. Amidst the light-hearted conversation, the robots, JoyBot and RoboGuide, continued to interact with them, their presence eliciting smiles and curiosity from both the doctors and the other customers.

JoyBot, with its cheerful demeanour, approached the table where Dr. Joshua Lee was seated. "Good day, Dr. Lee! How can I assist you today? A hot cup of coffee to keep you energised?"

Dr. Joshua Lee chuckled, appreciating the personalised attention. "Yes, JoyBot, a coffee sounds perfect. Surprise me with your specialty!"

As JoyBot skilfully prepared the coffee, Dr. Olivia Davis joined the conversation, sharing anecdotes of the robots' capabilities and their role in enhancing patient care. The doctors listened attentively, sipping their coffee, and nodding in awe.

Meanwhile, RoboGuide, with its polite and precise mannerisms, approached Dr. Rachel Patel, offering a selection of pastries. "Good afternoon, Dr. Patel. May I interest you in a freshly baked croissant or a delightful fruit tart?"

Dr. Rachel Patel smiled warmly. "I'll have the fruit tart, please. And thank you, RoboGuide, for the recommendation!"

The cafe buzzed with activity as the doctors enjoyed their coffee and engaged in lively conversations with the robots. The robots, programmed to provide friendly and helpful interactions, contributed to the relaxed and jovial atmosphere.

Dr. Samantha Aung, her eyes twinkling with amusement, turned to Dr. Sophie Bradley. "So, Dr. Bradley, do you think we'll have robot colleagues joining us on ward rounds someday?"

Dr. Sophie Bradley laughed heartily, imagining the scenario. "Who knows, Dr. Aung? With the advancements we've witnessed today, anything is possible!

The doctors continued to exchange stories and ideas, their laughter mingling with the hum of conversation and the soft whirring of the robots. The cafe provided a sanctuary for them to unwind, recharge, and appreciate the remarkable advancements of Quintomo Hospital.

As they savoured their coffee and enjoyed the company of their robotic assistants, the doctors marvelled at the seamless blend of technology and human interaction. It was a testament to the ongoing innovation and dedication of the hospital's team in harnessing the power of AI and robotics for the betterment of patient care.

In this moment of relaxation and camaraderie, the doctors found themselves inspired and invigorated, eager to continue their journey in this extraordinary era of medical advancement.

Sophie's First Emergency Code at Quintomo Hospital

Dr. Sophie Bradley quickly finished her lunch and said goodbye to her colleagues before receiving an urgent notification from Primeo about an emergency code. She immediately headed to the emergency department and grabbed one of the nearby emergency scooters, speeding through the designated paths and ramps to arrive at the scene of the emergency.

The patient, a middle-aged man, was being resuscitated by a team of doctors and nurses led by Dr. Jess Lewis, the emergency physician and Sophie's boss. A robot was performing CPR while the medical team worked furiously to stabilise the patient by ensuring that all the appropriate monitors were connected appropriately and establishing intravascular access to administer medications. The environment was chaotic with equipment strewn everywhere and the sound of alarms and machines beeping incessantly.

As Sophie introduced herself to the team, Primeo provided a loud and continuous status report of the resuscitation process, giving updates on the patient's condition including the statistical chance of survival and what action was next for the team following the Advanced Life Support (ALS) algorithm. Dr Lewis stood back to assess all the data points and gave the go ahead for every action being taken. With Primeo's guidance, the team was able to perform the necessary steps to increase the patient's survival chance from 50% to 90%.

Finally, the patient was revived, but he remained unconscious and was intubated before being transported to the intensive care unit (ICU) for further investigations. Prior to transfer Primeo gave a status report recounting the preceding event and patient's current status. The intensive care unit team had the opportunity to ask questions to the emergency department team, Primeo, and the patient's wife, Sarah. As Sophie and the team dispersed, the air was filled with a sense of relief, but also a hint of uncertainty as to the patient's ultimate prognosis.

Sophie couldn't help but wonder what kind of challenges she would face next as a resident at Quintomo Hospital. She knew that the experience would be intense, but she was excited for the opportunity to learn from some of the best medical professionals in the world. Meanwhile, the journey of this patient who was successfully resuscitated following a cardiac arrest was only beginning.

Chapter 3

THE CRITICAL CASE OF MATTHEW CLARK: A RACE AGAINST TIME

Matthew Clark woke up to a bright and sunny morning. He could hear his daughters giggling and chattering in their room. He smiled as he thought about the fun day they had planned. He got up from bed and walked into the girls' room, finding them bouncing up and down on their beds with excitement.

"Daddy, we're up! It's time to go to the park!" his 4-year-old daughter, Mia, exclaimed.

Matthew chuckled and scooped her up in his arms, kissing her on the cheek. He then turned to his 7-year-old daughter, Emma, who was already dressed and waiting by the door.

"Good morning, Dad," she said, giving him a big hug.

"Good morning, my love," he replied, kissing the top of her head.

He walked into the kitchen, where his wife, Sarah, was making breakfast. They chatted about their plans for the day as he poured himself a cup of coffee.

"So, what's on the agenda for today?" he asked.

"We thought we would take the girls to the park for a picnic and some fun in the sun," Sarah replied, flashing him a smile.

"That sounds perfect," Matthew said, taking a sip of his coffee.

After breakfast, they got dressed and packed a picnic basket with sandwiches, fruit, and drinks. They set off for the park, holding hands and chatting along the way.

Fun Family Time at a Quintomo Park

As Matthew and Sarah set up their picnic blanket, they couldn't help but take in the vibrant atmosphere of the park. Quintomo's futuristic cityscape provided the perfect backdrop, with sleek skyscrapers and colourful LED lights creating a mesmerising scene. The park itself was a testament to modernity, designed with state-of-the-art play equipment and innovative recreational spaces.

Mia and Emma wasted no time in exploring the playground, their laughter filling the air as they ran towards the swings and slides. Matthew and Sarah watched with joy in their eyes, grateful for this precious moment of family togetherness.

As the girls played, they encountered two other families who were also enjoying the park. The first family consisted of the Johnsons, a warm and friendly couple with their two energetic boys, Tyler, and Jake. The boys immediately struck up a playful conversation with Emma and Mia, their laughter merging with the sounds of the playground.

"Hey there! Looks like we've got some expert climbers over here!" Mr. Johnson exclaimed with a grin, pointing to Mia and Emma as they effortlessly manoeuvred through the climbing frames.

Sarah chuckled and replied, "Oh yes, they certainly have no fear when it comes to conquering heights!"

Meanwhile, nearby, the Parkers, a delightful family with a young girl named Lily, were enjoying a game of frisbee on the lush green grass. Lily's infectious giggles filled the air as she attempted to catch the frisbee with her small hands.

Matthew couldn't resist joining in the fun, offering to toss the frisbee for Lily. "You've got quite the arm, Lily! You're going to be a pro at this in no time!" he cheered.

The park itself was a marvel of modern design, featuring interactive sensory play areas, innovative climbing structures, and even a digital interactive water fountain that danced with colourful lights. The vibrant atmosphere created a sense of wonder and excitement for both children and adults alike.

As the afternoon sun cast a warm glow over the park, the families gathered on their picnic blankets, sharing stories and laughter. Matthew and Sarah couldn't help but feel a deep sense of gratitude for the beautiful memories they were creating with their daughters.

"Isn't it amazing how technology has enhanced our experiences in the park?" Sarah remarked, as she watched Mia and Emma engage in a friendly race with Tyler and Jake.

Matthew nodded, a proud smile on his face. "Absolutely! Quintomo has truly created a playground of the future. I'm grateful that our girls get to grow up in such an incredible city."

Matthew's Sudden Collapse

Suddenly, Matthew's face contorted in pain as he clutched his head.

"Are you okay?" Sarah asked, concern etched on her face.

"I don't know, I have a really bad headache," Matthew replied, gritting his teeth.

Sarah immediately became worried, as Matthew rarely complained about anything.

"Maybe we should head home and get some rest," she suggested.

But before they could even start packing up their things, Matthew collapsed to the ground, unconscious.

"Matty! Matty! Wake up!" Sarah cried out, shaking him frantically.

But he didn't respond, and she quickly rang for an ambulance. As she waited for them to arrive, she cradled his head in her lap, tears streaming down her face. The Johnsons and the Parkers were by their side providing support.

When the ambulance arrived, the paramedics quickly assessed Matthew's condition and took him to the emergency department of Quintomo Hospital.

Matthew Clark Suffers a Cardiac Arrest

As Dr. Jess Lewis, the emergency physician, began her assessment, Matthew suddenly had a cardiac arrest. This was the story recounted by Sarah, Matthew's wife in the emergency department after his resuscitation.

Dr. Ryan Baker, the intensivist in charge of Matthew Clark's care, assessed the patient with his resident doctors, Dr. Jane Smith and Dr. David Chang, and the intensive care unit (ICU) trainee, Dr. Emma Green. They were all standing outside the patient's room, looking at the MRI scan and discussing the results.

Dr. Baker turned to Dr. Smith, "Jane, what do you make of this? It looks like a bleed from a ruptured aneurysm."

Dr. Smith nodded, "Yes, that's what it looks like. It's not good news. We'll need to consult with the neurosurgery team and discuss treatment options."

Dr. Chang chimed in, "And his blood tests and ECG show evidence of a non-ST elevation myocardial infarction, probably a Type 2. We'll need to manage that as well."

Dr. Green spoke up, "What's the plan for his care?"

Dr. Baker replied, "For now, we'll stabilise him and monitor his condition. We'll consult with the neurosurgery and cardiology teams and come up with a plan of action."

Just then, the nurse unit manager, Emily, walked over to them. "Dr. Baker, we need to discuss the patient's care plan. Is there anything specific we need to be aware of?"

Dr. Baker turned to her, "Yes, we have a complex case here. The patient has suffered a ruptured aneurysm and a heart attack. We'll need to monitor his condition closely and be ready to act quickly if there are any changes. I'll make sure to keep you all updated on any new developments."

The nurses nodded, and Emily said, "We'll be sure to keep a close eye on him. Let us know if there's anything else we can do to help."

As soon as they stabilised the patient, Dr. Baker and his team went to the family meeting room to speak with Matthew's wife, Sarah, about his condition.

Dr Baker Discusses Matthew's Prognosis

"Mrs. Clark, I'm Dr. Baker, the Intensivist in charge of your husband's care," Dr. Baker introduced himself, shaking Sarah's hand.

Sarah looked anxious and worried as she replied, "How is my husband? Is he going to be okay?"

Dr. Baker sighed and took a seat beside her. "Your husband is currently stable, but he had a ruptured brain aneurysm that caused a bleed in his brain. We have put him on a ventilator to help him breathe and we are monitoring him closely. We also found evidence of a heart attack from the ECG and blood tests. He is currently unconscious, but we are doing everything we can to keep him stable."

Sarah's face went pale, and tears welled up in her eyes. "Is he going to make it? What are his chances?"

Dr. Baker hesitated for a moment before replying, "I won't sugarcoat it, Mrs. Clark. Your husband's condition is serious, and we can't predict the outcome at this point. We have called in an interventional cardiologist and a neurosurgeon to discuss his treatment options. We will keep you updated on his progress and let you know as soon as we have any new information."

Sarah nodded, still in shock from the news. "Can I see him?"

Dr. Baker nodded. "Of course, we will take you to see him in a moment. But I must warn you, he is currently unconscious and on a ventilator. We have a team of nurses and respiratory therapists monitoring him 24/7 to ensure his comfort and stability."

As they walked back to the ICU, Dr. Baker introduced Sarah to the nurses and respiratory therapists who were taking care of her husband. The nurse unit manager, Emily, took Sarah's hand and reassured her that they were doing everything they could to keep Matthew comfortable and stable.

Sarah nodded, tears streaming down her face as she saw her husband hooked up to various machines and monitors. She held his hand and whispered softly to him, hoping he could hear her.

Dr. Baker, meanwhile, called in the interventional cardiologist and neurosurgeon for consultation. The team discussed the best course of action for Matthew's treatment, weighing the risks and benefits of each procedure.

The interventional cardiologist, Dr. Amira Khan, suggested that the heart attack was likely secondary to the bleed in the brain, hence, it was best to first proceed with neurosurgical intervention if the patient remained stable. She would continue to review and arrange a coronary angiogram down the track. The neurosurgeon, Dr. Sally Brown, agreed to perform an endovascular coiling procedure to repair the aneurysm. An endovascular coiling procedure involves threading a thin tube through the blood vessels to the aneurysm site, then placing small coils within the aneurysm to promote blood clotting and prevent rupture. This was scheduled for the next morning. Dr. Baker thanked the consultants for their input and supported their suggested course of action.

As the day turned into night, Matthew remained unconscious but stable, his vital signs monitored closely by the ICU team. Sarah sat by his side, holding his hand, and praying for his recovery.

Dr. Baker checked in on them periodically, updating Sarah on the plans for her husband's treatment and offering her words of comfort and support.

It was a long and difficult night for Sarah and the ICU team, but they remained vigilant and focused on providing the best possible care for Matthew.

Matthew Clark Goes to Theatre

The next morning, Matthew Clark was taken to the operating theatre for his endovascular coiling procedure, accompanied by his wife who was still in a state of shock from the events of the previous day. Dr. Sally Brown, the neurosurgeon, was there to lead the team, along with a group of highly skilled surgical technicians and nurses.

As they prepped for the procedure, the AI medical assistant was activated, its robotic voice filling the room. "Good morning, Dr. Brown. This is Primeo, your AI surgical assistant for this procedure. I am ready to assist you in any way possible."

Dr. Brown smiled, "Good morning, Primeo. Let's get to work."

The team worked efficiently, with Dr. Brown manoeuvring the endovascular catheter through Matthew's arteries, using the AI assistance to guide her in real-time. Primeo was able to analyse the images in real-time, providing suggestions to Dr. Brown for optimizing the procedure. "You're slightly off-centre, Dr. Brown. Adjust the catheter slightly to the left," the AI assistance alerted her.

Dr. Brown quickly made the adjustments, and the team continued with the procedure. As they worked, the robotic arm positioned over the table helped hold steady the catheter, ensuring that the coil was placed exactly where it needed to be. The technology was state-of-the-art, and the team worked in seamless harmony with the robots, as if they had been working together for years.

Dr. Brown took a moment to speak with the robot, which she had affectionately named "Robo." "Robo, can you adjust the angle of the X-ray machine? I need a better view of the aneurysm," she requested.

"Of course, Dr. Brown. Adjusting angle now," the robot replied.

As the procedure progressed, Primeo provided real-time status updates, informing the team of Matthew's vital signs and other important parameters. The AI assistance also monitored his brain activity, ensuring that there were no further issues during the procedure.

After several hours, the procedure was complete. Matthew was wheeled back to the ICU, where he would continue his recovery. Dr. Brown spoke to his wife in the family meeting room, explaining that the procedure had gone well and that he would need to be monitored closely over the next few days. She also explained that a follow-up CT scan would be needed to confirm the success of the procedure.

Matthew Clark's case conference convened

Despite the successful procedure, Matthew remained unconscious five days after the surgery. His wife, family, and the medical team were growing increasingly concerned. Dr. Baker called for a case conference with the other specialists to discuss the patient's condition and decide on the next course of action.

In the conference room, Dr. Baker was joined by Dr. Khan, Dr. Brown, and a cardiology fellow, Dr. Cooper. They discussed the possible causes of Matthew's prolonged unconsciousness, which included the aneurysm rupture, the cardiac arrest, and the surgery itself.

Dr. Khan explained that the cardiac enzymes and echocardiogram were stable and that features of stress cardiomyopathy previously noted were unchanged. She felt it was unlikely that an acute cardiac problem was the cause of the prolonged unconsciousness, also considering his stable vital signs. Dr. Brown reported that the post-procedure imaging showed that the aneurysm was successfully coiled and there was no evidence of any new bleeding. However, there was some evidence of oedema in the surrounding brain tissue.

Dr. Cooper discussed the possibility of the patient suffering from a hypoxic-ischemic brain injury during the cardiac arrest, which could be causing the prolonged unconsciousness. He suggested that they perform a neurologic exam to evaluate the extent of the damage.

As they discussed the options, the AI medical assistant chimed in with suggestions and data analysis, providing real-time updates on the patient's vitals, lab results and statistics about his prognosis. The robot, named Robi, assisted in the discussion and offered tips on how to optimize the patient's care based on the available data.

After much deliberation, the team decided to perform a neurological exam to assess the extent of any brain damage, as well as a repeat CT scan to check for any changes in the brain tissue. They also decided to consult with a neurologist for input on the patient's case.

As the meeting ended, the team remained uncertain about the patient's prognosis. They had done everything in their power to provide the best possible care, but they were still unsure about what the future held for Matthew. The air of uncertainty hung heavy in the room as the team dispersed, each lost in their thoughts and worries about the patient's condition.

Chapter 4

DIFFICULT DECISIONS: NAVIGATING LIFE AND DEATH IN THE AGE OF TECHNOLOGY

Dr. Michael Wong stood at the patient's bedside, carefully reviewing patient information and imaging on the multiple large virtual screens enabled by his Primeo Glass. He discussed the patient's progress with the nursing staff and his team of medical students, residents, and specialist trainees. The patient, a 65-year-old man, had suffered a stroke and had undergone successful endovascular clot retrieval the day before. Endovascular clot retrieval is a procedure where doctors guide a small device through the blood vessels to the site of the clot causing the stroke, then capture and remove the clot to restore normal blood flow to the brain. With no residual deficits and appropriate medication, the patient was ready to be discharged.

"Well, Mr. Smith, I'm happy to say that you're doing very well," Dr. Wong said with a smile. "Your recovery has been remarkable."

"Thank you, Doctor," Mr. Smith replied gratefully.

Dr. Wong turned to his team of residents and trainees. "So, let's review some key points about stroke management. When is thrombolysis indicated?"

One of the residents quickly responded, "Within 4.5 hours of symptom onset, if there are no contraindications."

"Good," Dr. Wong nodded. "And what about clot retrieval?"

Another resident spoke up, "It's recommended within 24 hours of symptom onset for patients with proximal occlusion who meet certain criteria."

"Excellent," Dr. Wong praised. "And what role does perfusion imaging play in stroke management?"

There was a brief pause before one of the trainees spoke up, "Perfusion imaging can help identify salvageable brain tissue and guide decision-making about reperfusion therapies even beyond the 4.5-hour window."

"Exactly," Dr. Wong said, impressed. "Now let's all go grab some coffee at the hospital cafe and discuss some interesting cases."

The Robo-Powered Cafe

The team arrived at the Robo-powered cafe for their coffee break. Dr. Michael Wong sat down and received his coffee, which was prepared and served by a robot. Two trainees who were regulars at the cafe also sat down with their coffee. The rest of the team went to order their coffee from the robo-stand.

As they sat down to enjoy their coffee, the medical student on the team, Luke, couldn't stop talking about the cool experience of being served by a robot. The team members chimed in and shared stories and anecdotes of their experiences with the advanced technology at Quintomo Hospital, including Primeo and other robots.

The Robo-cafe used facial recognition to match the staff's identity and match their profile, making for express and personalised service. For those who didn't sign up for express service, the process was still quick and easy – they simply had to place their order at one of the many robo-stands, swipe their payment card, and wait for their coffee to be dispensed.

Matthew Clark's Case Discussed Over Coffee

Dr. Wong, who was known for his clinical brilliance but also for his lack of interpersonal skills, remained mostly quiet during the conversation, sipping his coffee, and listening to his team members share their experiences.

Dr. Wong sipped his coffee as he listened to the chatter around him, occasionally nodding in agreement. As the conversation died down, he cleared his throat and spoke up. "You know, I'm heading to the ICU shortly to review a patient who had a cardiac arrest. But I think after a few days of no improvement, it's time to consider stopping active treatment. Prolonging their life is just prolonging their suffering," he said with conviction.

The other team members looked at him in surprise, the junior doctors saw things differently. "But Dr. Wong, shouldn't we give the patient more time to recover?" asked one of the trainees.

Dr. Wong sighed. "I've seen too many cases where active treatment was continued for too long, even when there was no hope for recovery. It's just prolonging their pain and suffering. And don't get me started on the AI, some doctors in this hospital seem to think it's omnipotent," he replied, shaking his head.

The team members exchanged glances, not sure what to say. Dr. Wong finished his coffee and stood up. "I'll see you all in the clinic this afternoon," he said before heading off to the ICU.

The team members watched him go, still processing his strong opinions.

Matthew Clark's Case Conference Convenes

Dr. Michael Wong walked into the conference room where the intensivist, Dr. Ryan Baker, interventional cardiologist, Dr Amira Khan and neurosurgeon, Dr. Sally Brown were already seated. He had just finished the neurological examination on the patient and conducted an electroencephalogram (EEG), and his assessment was that the patient had severe hypoxic brain injury and was unlikely to survive. He recommended a discussion with the family in view of palliation.

As soon as Dr. Wong presented his opinion, the neurosurgeon, Dr. Brown, spoke up, "I'm not sure I agree with you, Michael. This patient is young and has a chance of recovering. I suggest we wait for a week before reassessing to make a final decision."

Dr. Wong bristled at the neurosurgeon's suggestion. "I understand your concern for the patient, but we must be realistic about the prognosis. The EEG results are conclusive, and the patient has little to no chance of recovery. Prolonging life-support measures will only prolong the inevitable."

The intensivist, Dr. Baker, chimed in, "I agree with Dr. Wong. We should consider palliative care and focus on the patient's comfort."

The interventional cardiologist, Dr. Khan, remained neutral, "I hear both sides of the argument, but ultimately, the family should be involved in the decision-making process."

The room fell into silence as each specialist pondered the options. Finally, the neurosurgeon spoke up again, "Okay, I understand your point of view, Michael. But please, give me and the family some time to process this information. We'll reassess the patient in a week and make a decision then."

Dr. Wong reluctantly agreed, "Fine, we can reassess in a week. But I hope that we can all keep the patient's best interests in mind."

The case conference ended, and each specialist went their separate ways. Dr. Wong couldn't help but feel frustrated with the neurosurgeon's opinion. He asked Primeo to schedule a follow-up appointment for next week and note his recommendation for a palliative approach.

As he exited the conference room, he was met with a group of nurses who had been waiting outside. They bombarded him with questions about the patient's condition, and he patiently answered each one, giving them updates on the case.

He then requested a status report from Primeo about his forthcoming clinic before heading back to his office. He needed to clear his head before seeing his trainees in the clinic that afternoon.

One week later, a family meeting to decide Matthew Clark's fate

One week later, Matthew Clark had made no improvement and the specialists decided to discuss with his family about turning off the ventilator. As the intensivist, Dr. Ryan Baker, and his team walked into the room, they could feel the tension in the air. The patient's wife, Sarah, was sitting beside her mother-in-law, June and sister-in-law, Audrey, all with tears in their eyes. Dr. Baker took a deep breath and began, "I want to thank you all for coming today. As you know, your husband has suffered a cardiac arrest and has been in the ICU for the past couple of weeks. Unfortunately, his brain was deprived of oxygen for a prolonged period of time, and despite our best efforts, he has not shown any signs of improvement."

Sarah, Matthew Clark's wife, interrupted, "But the neurosurgeon said that he could have more time. Why are you giving up on him so soon?"

Dr. Baker replied calmly, "I understand how difficult this is for you, and we are not giving up on him. We are simply proposing a change in the focus of care towards providing comfort and allowing nature to take its course."

Sarah shook her head vehemently, "No! He is a fighter; he wouldn't want us to give up on him like this. He loves life, he's never been sick before, and he did all the right things."

Dr. Baker looked at the patient's chart and said, "I understand that he has been healthy, but unfortunately, this was a sudden and catastrophic event. We have done everything we can, but the reality is that he is not going to recover from this."

Sarah looked at Dr. Baker with anger and frustration, "But how can you say that? The cardiologist didn't seem too worried about his heart, and the neurosurgery went well. How can you be saying that you want to turn off the ventilator and let him die?"

Dr. Baker replied with empathy, "I know this is hard for you to understand, but the reality is that his brain has been severely damaged. The interventions we have done so far were to try to give him the best chance of recovery, but unfortunately, they have not worked."

Sarah continued to cry and plead, "Please, can't you try something else? Can't you give him more time? He deserves a chance!"

Dr. Baker responded with compassion, "We have already given him the best chance we could, and unfortunately, it hasn't worked. We are not giving up on him, but we need to shift our focus towards providing him with comfort and making sure he is not suffering. We will do everything we can to support him and your family during this difficult time."

The tension in the room was palpable as Sarah Clark and her family struggled to come to terms with the proposed change in focus of care. Dr. Baker and his team did their best to offer support and comfort to the family, but it was clear that this was going to be a long and difficult process for everyone involved.

Dr. Baker took a deep breath before speaking. "I know this is a difficult time for all of you, and I understand that you are looking for answers. I want you to know that we have done everything possible to help your husband, but sometimes, even with the best medical care, the outcome is not what we had hoped for."

Sarah looked up at Dr. Baker. "But is there anything else that can be done? Anything at all, even if it's experimental?"

Dr. Baker hesitated before answering. "We could explore other treatment options, but I must be honest with you. Given your husband's condition, it is unlikely that any experimental treatments would be effective."

The patient's mother, June, interjected. "But this hospital is known for its advanced technology and AI, isn't it? Maybe there's something they can do."

The patient's sister, Audrey, nodded in agreement. "Can we get a second opinion from the AI department?"

Dr. Ryan Baker, the intensivist, took a moment to consider their request. "I will arrange for a consultation with the AI team, but I cannot promise anything. We will do everything in our power to help your husband, but we also need to be realistic about his condition."

The family nodded, and the meeting was adjourned. As they left the room, Dr. Baker and his team could hear the patient's wife and mother quietly weeping. It was clear that this was going to be a long and difficult process for everyone involved.

Chapter 5

NAVIGATING SKEPTICISM: PROF. BILL STONE'S JOURNEY TO BRING AI TO HEALTHCARE

Prof. Bill Stone, the Chief AI Officer at Quintomo Hospital, is a 55-year-old Australian-born medical pioneer hailing from the northern beaches suburb of Sydney. Raised in an environment steeped in medical expertise and analytical thinking, his parents played a significant role in shaping his aspirations. His father, a renowned radiologist, and his mother, an analyst at McKinsey, instilled in him a deep passion for both medicine and analytical problem-solving. With their unwavering support, Prof. Stone's ambition to make a difference in the field was ignited and continued to drive his relentless pursuit of innovation.

After completing his medical education at the esteemed University of Melbourne, Prof. Stone set his sights on the United

States for specialist training. He embarked on a transformative journey, honing his skills at prestigious institutions such as Harvard Medical School, where he completed his residency in radiology. It was during his time at Stanford Medical School that Prof. Stone immersed himself in the groundbreaking field of AI in healthcare, becoming one of the early adopters and pioneers of this emerging discipline. He furthered his knowledge by obtaining a Ph.D. in deep learning, focusing on the development of prediction models and quality control systems.

Family Life and Personal Interests

While at Stanford, Prof. Stone's path crossed with that of his future wife, Dr. Angela Fuller, an American data scientist and researcher. Their shared passion for technology and innovation sparked an immediate connection, leading to a profound partnership that extended beyond their personal lives. Together, they navigate the challenges and triumphs of the ever-evolving landscape of healthcare and artificial intelligence. They are blessed with two beautiful children, Shaun and Abby, who bring boundless joy and inspiration to their lives.

Prof. Stone's passion for innovation is not confined to the walls of Quintomo Hospital. In his personal life, he finds solace in nature, embarking on exhilarating hikes that serve as a source of inspiration for his work. He is an avid reader, delving into science fiction novels and thought-provoking literature on artificial intelligence and machine learning, constantly seeking new insights and ideas to push the boundaries of medical technology.

Despite his remarkable achievements, Prof. Stone remains grounded and approachable, known for his collaborative spirit and unwavering dedication to his colleagues.

Prof. Stone's Role at Quintomo Hospital

As the Chief AI Officer at Quintomo Hospital, Prof. Bill Stone continues to spearhead transformative advancements in healthcare. His vision, expertise, and unyielding commitment to patient care have positioned him as a driving force in the integration of AI and technology within the medical field. With his keen intellect, unwavering determination, and unquenchable thirst for progress, Prof. Stone is revolutionising the way healthcare is delivered at Quintomo Hospital and beyond.

A Day in the Life of Prof. Bill Stone

Prof. Bill Stone sat at the breakfast table with his family in their beautiful beachfront home. The sun was just beginning to rise, casting a warm glow through the windows. His wife, Angela, was pouring coffee while their two children, Shaun and Abby, were getting ready for school.

"So, what's on the agenda for today?" Angela asked, handing him a cup of coffee.

"I'm giving a presentation at the grand rounds this afternoon," Bill replied, taking a sip of coffee. "I'll be talking about the latest developments in AI and healthcare."

"That sounds exciting," Angela said. "Remember when you were still at Stanford and how hard it was to implement change in the early days of AI in healthcare? It's amazing how far you've come."

Bill nodded. "Yes, those were tough times, but we're making progress. It's a long journey, but we'll get there."

Angela sipped her coffee thoughtfully. "I miss our old friends and some aspects of our lives in Stanford, but I really like it here now in Quintomo. The community is great, and I love how close we are to the ocean. Plus, the kids are happy here."

Bill smiled at her. "I'm glad you're happy here. We're lucky to live in such a beautiful place. I'm glad to be back in Australia to be closer to my aging parents."

Shaun and Abby came to the table, grabbing their backpacks and sitting down to eat. Bill took a moment to appreciate his family, grateful for their support and love.

"Remember, kids, make sure you're respectful and kind to everyone you meet today," he said, kissing them both on the forehead.

As they finished up breakfast and headed out the door, Bill took a moment to soak in the beautiful view of the ocean. He felt fortunate to have such a supportive family and a fulfilling career, even if it wasn't without its challenges.

Prof. Stone's Workday Begins

Prof. Bill Stone arrived at his desk, put on his Augmented Reality (AR) glasses and asked Primeo about his schedule for the day. "Good morning, Primeo. Could you please give me a rundown of my schedule today?" he asked.

"Good morning, Prof. Stone. You have a meeting with the department heads at 9am, followed by a meeting with the CTO at 10am. Then, at 11am, you have patient consultations scheduled. You also have a meeting with 10 unit heads of various aspects of AI research at 12pm. At 1pm, you have a meeting with the CEO, and then you are scheduled to present at Grand Rounds at 2pm," Primeo replied.

"Thank you, Primeo. Can you give me more details about the department head's meeting and the last meeting's action points?" Prof. Stone asked.

"Of course, Prof. Stone. The department head's meeting will be discussing further collaborations on some protocols they are developing, and they are seeking funding from management for some new AI tools. The action points from the last meeting were to finalise the protocols and submit the funding proposal," Primeo answered.

"Great. And how about the meeting with the CTO?" Prof. Stone inquired.

"The meeting with the CTO will be to discuss the upcoming system upgrade and some privacy recommendations from the ethics committee," Primeo replied.

Prof. Stone nodded, then asked about his patient consultations. Primeo informed him that he had five referrals to review by 11am that morning. Prof. Stone reviewed his schedule again and realised that he needed to reschedule a few meetings to allow for adequate preparation for his Grand Rounds presentation.

"Primeo, please reschedule my meeting with the CEO to 12:30pm instead. I'll need some extra time to prepare for my Grand Rounds presentation at 2pm," he said.

"Understood, Prof. Stone. The meeting has been rescheduled as requested," Primeo replied.

Prof. Stone then began planning for his upcoming meetings and tasks of the day.

Prof. Stone's Patient Consultations

Later, he reviewed the status reports of various medical cases referred to him. He asks Primeo to display the imaging results for the cases he was reviewing.

"For this case, let's review the imaging results for patient Mary Johnson," he says, looking at the screen. "What does the MRI show?"

"Displaying the images now, Prof. Stone," Primeo replies, projecting the images in front of him.

Prof. Stone studies the images carefully, noting the areas of concern. "This looks like it could be an early-stage tumour," he says. "Let's consult with the oncology team on this one."

As he moves on to the next case, he notices a flagged recommendation from the AI system.

"Primeo, what's the reason for the flag on this case?" he asks.

"The AI system gave a recommendation that was deemed inappropriate by the treating specialist," Primeo responds.

Prof. Stone frowns. "That's concerning. Let's expedite a review of the reason for the erroneous recommendation by the Quality and Safety team. They need to investigate the incident, report on the root cause, and make recommendations to prevent similar errors in the future."

As he reviews the final case, which is about Matthew Clark, who is in the intensive care unit (ICU), Prof. Stone's expression turns grave.

"Matthew's condition has deteriorated significantly," he says. "We need to discuss the possibility of enrolling him in a clinical trial for experimental treatments. Please reach out to the research team to discuss their recommendations." Prof. Stone took a deep breath and continued reviewing the cases, determined to provide the best possible care for his patients.

Prof. Stone's Grand Rounds Presentation

It was now time for the Grand Rounds presentation. Prof. Bill Stone stood at the podium, looking out over the crowded room. The hospital's Grand Rounds was always a popular event, and he was pleased to see so many familiar faces in the audience.

"Good morning, everyone," he began, smiling at the residents, specialists, and researchers in attendance. "Today, I want to talk to you about the exciting developments in artificial intelligence and how they're transforming the field of healthcare."

As he spoke, he saw the audience nodding in agreement, some taking notes and others asking questions. Prof. Stone gave an overview of the principles of machine learning, deep learning, natural language processing, nanotechnology, and sensors that underpinned some of the technology they use at the hospital.

He went on to discuss the results of recent studies, including those that they were a part of, on the use of AI in phlebotomy, drug administration, and precision medicine.

"By using AI to analyse large datasets, we can better understand a patient's genetic makeup, their medical history, and the specifics of their illness," he said. "This enables us to deliver targeted treatments that are tailored to the individual, improving outcomes and reducing costs."

Prof. Stone finished his presentation to thunderous applause, and the audience began to disperse. As he stepped down from the podium, he was greeted by some of the hospital's residents.

"That was incredible, Prof. Stone," one of them said. "I never realised just how powerful AI could be in healthcare."

"I'm glad I could shed some light on the subject," Prof. Stone replied with a smile.

The other specialists in attendance approached Prof. Stone, congratulating him on the informative and engaging presentation.

"I'm excited to see what the future holds for AI in healthcare," one of them said.

Prof. Stone nodded in agreement. "We're only scratching the surface of what's possible," he said. "But I have no doubt that AI will continue to transform healthcare and improve patient outcomes in ways we can't even imagine yet."

As he made his way back to his office, Prof. Stone felt energised and inspired. He knew that there was still much work to be done, but he was proud of the progress they had made so far, and confident in the future of AI in healthcare.

Prof. Stone discusses Matthew Clark's case with treating clinicians

After the grand rounds, Prof. Stone had a brief chat with the team of doctors looking after Matthew Clark - Dr Amira Khan, Interventional Cardiologist, Dr Sally Brown, Neurosurgeon, Dr Michael Wong, Neurologist, and Dr Ryan Baker, Intensivist. They approached him for a chat in front of the auditorium. He had reviewed Matthew's case and acknowledged it was a hard one. However, he just heard back from the research team before coming to the grand rounds. After reviewing their algorithm's recommendations, they favoured enrolling him in 2 clinical trials - one that involved neural data transmission via electrodes implanted into the brain to evaluate prognosis and assess the extent of reversibility of brain injury, and the second involving stem cell treatment to improve outcomes in severe brain injury.

Dr Khan spoke up first. "We really appreciate your input, Prof. Stone. We were at a loss as to what to do next with Matthew, and these clinical trials could be just what he needs."

Dr Brown nodded in agreement. "Yes, we've done everything we can with surgery and medication, but it's not enough. We need to explore new avenues to help him."

Dr Baker, added his support. "I agree. We need to explore every option available to us."

Dr Amira Khan, Dr Sally Brown, and Dr Ryan Baker eagerly await Dr Michael Wong's response, but he remains silent for a moment, lost in thought. Finally, he speaks up, his tone sceptical.

"I understand the importance of clinical trials, but we need to consider the risks involved," Dr Wong says. "These treatments are still in their infancy, and we don't know the long-term effects they may have on the patient."

Dr Khan responds, "We know the risks, Michael, but we also know the potential benefits. These clinical trials offer a chance for Matthew to recover, to have a better quality of life. It's worth considering."

Dr Brown nods in agreement. "I've seen cases where patients have made remarkable recoveries with these treatments. We owe it to Matthew to give him every chance we can."

Dr Wong takes a deep breath, considering their arguments. "I understand your point of view," he says. "I'm just concerned about the risks. But ultimately, this decision is up to Matthew's family. If they want to try the clinical trials, we'll support them."

Prof. Stone nods, "I agree. It's ultimately the family's decision, but we will ensure that they are well-informed of the risks and benefits before making any decisions."

The doctors agree, knowing that this will be a difficult decision for Matthew's family. Dr Baker proposed that Prof. Stone meet with Matthew's family the next morning at 10am. Primeo says, "You have no conflicts." Prof. Stone replies, "Lock it in."

As they dispersed, Prof. Stone reflected on what a busy but productive day it had been thus far. He headed back to the office to catch up on their multiple research projects and clearing up his inbox of status reports.

Chapter 6

HOPE ON THE HORIZON

Prof. Stone and the team of doctors sat with Matthew Clark's wife, Sarah, mother, June, and sister, Audrey, to discuss the clinical trials that could be done for Matthew's case. They went over the potential risks and the benefits of each trial, explaining them in detail. As the discussion went on, the tension in the room began to rise. The family members were anxious and had many questions that needed answers.

"What are the chances that this will work?" Sarah asked, her voice trembling.

Prof. Stone explained that while there was no guarantee, the clinical trials had shown promising results in similar cases. However, there was still a chance that the treatment might not work, and that Matthew may ultimately die.

June and Audrey looked at each other, tears welling up in their eyes. They couldn't bear the thought of losing him, but they knew that they had to try everything possible to save his life.

After a few moments of silence, Sarah spoke up. "We have to try. If there is even a chance that this could save him, we have to take it."

The team of doctors nodded in agreement, relieved that they had finally gained the family's consent. They handed Matthew's wife a tablet, and she signed the informed consent forms.

As the family left the room, the doctors exchanged glances, each one aware of the gravity of the situation. They knew that the clinical trials were not a sure thing, but they also knew that they had to try everything possible to save Matthew's life. The electrode brain implant was scheduled for later that afternoon.

Sefo's Journey from Fiji to Quintomo

Sefo, an 8-year-old boy from Fiji, was accompanied by his loving parents, father Tomasi and mother Adi, as they ventured into the bustling city of Quintomo for the first time. The family had arrived the day before their appointment with Dr. Sam London, excited and curious about the new city and the possibilities it held for Sefo's treatment.

Tomasi, Sefo's father, was a high school teacher in Suva, teaching biology and chemistry, while Adi, Sefo's mother, worked as a nurse at the Suva Teaching Hospital. They had been referred to Dr. London's clinic after Sefo was diagnosed with Ewing sarcoma, and the Fiji government had graciously funded their trip and medical expenses.

As a middle-class family experiencing their first time overseas, the wonders of Quintomo captivated them. They marvelled at the city's advanced technology, from self-driving trains that whisked them through the cityscape to flying self-driving taxis that soared above the streets, providing breathtaking views of the metropolis below. Even the self-driving cars that navigated the roads seemed like something out of a science fiction novel.

Walking through the city, they were amazed to see humanoid robots interacting with people, performing various tasks, and providing assistance. In the parks, they observed robots accompanying elderly citizens on their walks, offering a helping hand or a friendly conversation. It was a heartwarming sight, showcasing the seamless integration of technology and humanity.

One day, the family decided to visit an ultramodern, futuristic mall, where they were greeted by a spectacular display of interactive holograms and AI-guided shopping experiences. They browsed through the shops, trying on clothes that were chosen for them by intelligent algorithms, which considered their personal preferences and unique style. In one store, they even witnessed a robot fashion show, with the latest trends being modelled by lifelike mannequins that moved gracefully down the runway.

Feeling hungry, they decided to try a highly recommended restaurant known for its robo-service. As they entered, they were welcomed by a smiling robotic host who escorted them to their table. The ambiance was cozy and elegant, with soft lighting and soothing music filling the air. They browsed through a digital menu, which was projected onto the table, and selected their dishes with a simple touch.

As they waited for their meal, they observed the robotic waitstaff gliding through the restaurant, efficiently serving diners with grace and precision. It was a mesmerising sight, and they could hardly contain their excitement when their own robo-waiter arrived with their meal. The food was delicious, and the presentation was impeccable, making for an unforgettable dining experience.

Throughout their stay, Tomasi, Adi, and Sefo made the most of their time together, exploring the city's wonders and embracing its vibrant culture. Their experiences in Quintomo brought them closer as a family and gave them hope for a brighter future.

As they prepared for Sefo's appointment with Dr. London, they felt a sense of optimism. They knew that the journey ahead would be challenging, but they were grateful for the expertise of Dr. London and his team, and the cutting-edge medical technology that could potentially save their son's life. With love, determination, and the wonders of Quintomo by their side, they faced the future together, ready to conquer whatever challenges it held.

At the Clinic with Dr. London

Dr. Sam London, a respected paediatric oncologist, walked into his clinic, a bright smile on his face. His nurse, Sarah, and trainee registrar, Dr. Omar Khan, were already waiting for him, busy preparing for the day ahead.

"Good morning, Sarah, Omar. How are we today?" Dr. London asked, looking at his team.

"Morning, Dr. London. We're doing well, thank you," Nurse Sarah replied with a warm smile.

"Excited to learn and help, Dr. London," Omar chimed in.

"That's the spirit, Omar," Dr. London said encouragingly. "Now, let's go over the patients we'll be seeing this morning."

Primeo, their AI assistant, projected the status report of the four children scheduled for the day. The children, aged between

4 and 12, all had cancer cases that could benefit from molecular treatment after genetic testing and precision medicine algorithms.

"Primeo, please provide a brief overview of the cases we're seeing today," Dr. London requested.

"Certainly, Dr. London," Primeo replied. "Our first patient is a 6-year-old girl named Emily. She has neuroblastoma with an ALK mutation. Genetic testing revealed a potential sensitivity to targeted ALK inhibitors."

"Interesting. This could be a promising treatment option for Emily," Dr. London said, looking at Sarah and Omar. "Next?"

"Second is Jack, a 12-year-old boy with acute lymphoblastic leukaemia (ALL) and a Philadelphia chromosome-positive status. His case would benefit from targeted therapy using tyrosine kinase inhibitors, in combination with chemotherapy," Primeo continued.

"Challenging case, but we have a plan," Dr. London noted. "Go on, Primeo."

"Patient three is Isabella, a 4-year-old girl with a BRAF-mutated low-grade glioma. Molecular profiling suggests she may respond well to BRAF inhibitors, which are currently in clinical trials for paediatric patients with this specific mutation," Primeo said.

Dr. London nodded. "We'll have to discuss her case further and evaluate if she's eligible for a trial. And the last patient?"

"Lastly, we have Sefo, an 8-year-old boy from Fiji with Ewing sarcoma. Genetic testing identified a EWSR1-FLI1 fusion gene, which is a potential target for novel small molecule inhibitors that are being developed," Primeo finished.

"Four challenging cases, indeed," Dr. London said. "But we have the tools and the expertise to make a difference in their lives. Let's review their charts in detail and prepare for their consultations."

Nurse Sarah and Dr. Omar Khan nodded in agreement, ready to assist Dr. London in bringing the best possible care to these young patients. With the power of genetic testing and precision medicine, they were hopeful that they could provide these children with a fighting chance against their cancers.

Sefo's Consultation

Sefo and his parents, Tomasi and Adi, entered Dr. London's ultramodern clinic room with a mixture of excitement and trepidation. The room was spacious, filled with natural light from floor-to-ceiling windows that offered a stunning view of the city. The walls were adorned with holographic art, and the furniture was sleek and ergonomic. State-of-the-art diagnostic equipment was discreetly integrated into the room, making the space feel more like a sophisticated lounge than a traditional medical examination room.

As they settled in, they were greeted by the warm smiles of Dr. London, Nurse Sarah, and Dr. Omar Khan, the paediatric oncology trainee. Dr. London's demeanour was kind and reassuring, immediately putting the family at ease.

"Welcome, Tomasi, Adi, and Sefo," Dr. London began. "I'm glad you could make it here, and I hope your time in our city has been enjoyable so far."

Tomasi replied, "Yes, Dr. London, we've had a wonderful time exploring Quintomo and experiencing all the incredible technology. Thank you for having us."

"Now," Dr. London continued, "let's discuss Sefo's case in detail. We have already reviewed the blood tests and imaging results you sent us. Primeo, our AI assistant, has been analysing the data to identify the best treatment options based on Sefo's unique genetic profile."

Turning to Sefo, Dr. London asked, "How are you feeling today, Sefo?"

Sefo hesitated for a moment, then responded with a small smile, "I'm okay, Dr. London. A bit nervous, but excited too."

"That's perfectly normal, Sefo," Dr. London assured him. "We're here to help you, and we'll do everything we can to make sure you're comfortable and well-informed."

Throughout the conversation, Primeo was an active participant, displaying relevant information on the monitors and chiming in when necessary. After analysing Sefo's case, Primeo presented the family with several potential clinical trials that focused on molecular treatments for Ewing's sarcoma. The AI outlined options that included targeted therapy drugs and immunotherapy approaches, such as CAR T-cell therapy.

"Based on Sefo's specific genetic profile, I recommend discontinuing the chemotherapy he received in Fiji," Primeo advised. "Instead, I suggest enrolling him in a clinical trial that is investigating CAR T-cell therapy. CAR T-cell therapy is a treatment where doctors take a patient's own immune cells, genetically modify them in a lab to make them better at spotting and destroying cancer cells, and then put them back into the patient's body to fight the disease. This immunotherapy approach has shown promising results in other types of cancer and may offer Sefo the best chance for a successful outcome."

Dr. London nodded in agreement, elaborating on the potential benefits of CAR T-cell therapy and how it could specifically target Sefo's cancer cells. He explained the process, risks, and expectations to Tomasi and Adi, ensuring they had a thorough understanding of the treatment plan.

After discussing the treatment options, Sefo's parents felt confident in Dr. London's recommendation and agreed to proceed with the CAR T-cell therapy. They signed the informed consent form, acknowledging the potential risks and benefits of the treatment.

As the meeting concluded, the family felt a sense of optimism and relief. They knew they were in good hands with Dr. London and his team, who were dedicated to providing the best possible care for Sefo. As they left the clinic, they felt a renewed hope for Sefo's future, bolstered by the cutting-edge technology and compassionate care that surrounded them.

Matthew Clark's Brain Implant Assessment

Matthew Clark returned to the ICU after the successful implantation of the brain electrode. His family was by his side as Prof. Stone arrived with a team from the AI unit. Dr. Brown, Dr. Wong, Dr. Khan, and Dr. Baker were all there to assist in the assessment.

"Good afternoon, Mr. Clark. I hope you're feeling okay," Prof. Stone said as he approached the bed.

Matthew's family looked on anxiously, hoping for good news. Prof. Stone continued, "What we're about to do is connect our equipment to the sensors on your head. We hope to interpret the neural signals coming through to ascertain if you have irreversible brain damage. If the signals can be translated into words, that would be the best-case scenario."

Prof. Stone turned to the AI team and nodded, and they began to power up their equipment. The atmosphere in the room was tense, and Matthew's family was visibly anxious.

"Doctor, what are the chances of success?" Sarah, Matthew's wife, asked.

Prof. Stone took a deep breath before responding, "I have to be honest with you. This is a very difficult case, and the odds are not in our favour. However, we believe that this is worth trying, and we will do everything we can to help your husband."

Matthew's sister, Audrey, spoke up, "Can you explain how this works in layman's terms?"

Prof. Stone nodded, "We're essentially trying to read the signals that your brother's brain is sending out. We'll be looking for patterns that could indicate damage or recovery. It's a very complex process, but we have the best team working on it."

With that, the AI team began their assessment, connecting their equipment to the sensors on Matthew's head. The room fell silent as they analysed the signals coming through, searching for any glimmer of hope.

Prof. Stone looked at the screen in amazement as the signals appeared. He turned to the family, a smile spreading across his face. "This is incredible," he said, "We're getting neural signals from Matthew's brain. It's a great sign that he's able to communicate with us."

Matthew's wife and family were overwhelmed with joy and relief. They leaned over to him, hugging and kissing him as tears filled their eyes.

Prof. Stone explained that they were in the process of translating them into clearer messages. After adjusting the settings of the equipment and pressing some buttons, the following words appeared on the screen, which was visible to all: "Hi Sarah, hi everyone. I'm here but can't talk much." The atmosphere erupted with excitement, and there was even more animation from Matthew's family.

Prof. Stone explained what happens next, affirming the positive note but tried not to give them false hope as they were not completely out of the woods. He said that based on the result, they could now proceed with the stem cell therapy clinical trial and wrapped up the meeting.

CHAPTER 7

NEW GENERATION NURSES

At Quintomo Hospital, the nursing staff work in a high-tech environment that has revolutionised patient care. Robots now handle all patient transfers, freeing up the nursing staff to focus on direct patient care. Patients unable to mobilise are transferred to motorised, voice-activated self-driving wheelchairs. The patient care robots are always on standby, ready to assist as needed.

The nursing staff wear designated scrubs and use a Primeo device pinned by magnet to their left chest, just like the doctors. They also wear an earpiece for communication and use Primeo Glass for their virtual workstation, review of patient data, and live feed of the patient. The technology has improved the efficiency of patient care, resulting in fewer errors and increased patient satisfaction.

The use of sensors for patient monitoring has also made a huge difference. With more granular data such as the level of hydration, the severity of pain, the quality of sleep, frailty score etc obtained in addition to vital signs and cardiac rhythm. The nurses can quickly identify changes in a patient's condition and take appropriate action. This has led to improved patient outcomes and has helped prevent adverse events.

Despite the benefits of the technology, some nurses expressed concerns about losing the personal touch in patient care. They worried that the use of robots and other technology will lead to less face-to-face interaction with patients, potentially leading to a loss of empathy and connection. However, the nursing leadership reassured them that the technology is meant to supplement their work, not replace it.

Nurse Unit Managers Meet

The nursing director, Jen Nguyen, took her seat at the head of the conference table in the boardroom, as the nurse unit managers filed in and took their seats. They greeted each other warmly, chatting and laughing as they settled in. Jen, a seasoned nurse leader, began the meeting by acknowledging the hard work of the unit managers and their teams.

After discussing some housekeeping items, Jen moved onto the agenda topics, starting with the use of technology in nursing. She asked the group for feedback on the use of the latest version of the electronic health records, and whether they were seeing any issues or concerns with their use. The unit managers offered their perspectives, with some expressing frustration over the steep learning curve, while others noted that they appreciated the improvements in the upgrade.

Jen Nguyen, director of nursing, called upon Dr. Anu Patel, director of nursing AI research and implementation, for an update.

Dr. Patel replied, "Thank you, Jen. I wanted to update everyone on the impact that AI and robotics have had on our hospital so far. Data from last quarter shows ongoing significant reduction in work-related injuries, falls, drug errors, and improvement in nurse-to-patient ratios. The use of technology has certainly improved patient safety and quality of care."

The nurse overseeing patient flow, Mary Johnson, interjected excitedly, "And let me add, bed occupancy has been on average at 80%. We have seen a huge improvement since the implementation of the algorithm for bed allocation. It's such a relief not to have to deal with constant pressure to find beds for patients when the hospital is always at full capacity like in my previous hospital. It's all on autopilot!"

Dr Patel smiled, "Thank you, Mary. That's great to hear. We have seen significant improvements in patient flow since the implementation of the algorithm."

The nurse responsible for audits, Jessica Thompson, then presented the results of the recent audits, highlighting areas where nurses could improve their performance and patient care. "Overall, we've had mixed results in our audits. We've seen areas where nurses have excelled, such as timely documentation and adherence to medication protocols. However, there are areas where we need to improve, such as hand hygiene and triage of status reports sent by Primeo."

The meeting then turned to the discussion of two serious incident reports:

1. "A nurse failed to properly set up the sensor monitoring system on a patient with a history of heart issues. The sensor did not detect the patient's worsening condition, leading to a cardiac arrest."
2. "A patient with mobility issues was transferred by a robot to another part of the hospital, but the robot accidentally hit a wall, causing the patient to fall and sustain serious injuries." They deliberated on the cases and referred them for root-cause analysis with a report of recommendations to be reviewed in a future meeting.

Telemedicine Clinic

Dr. Kirra Nakata is accompanied by Dr. Rachel Patel, the intern, Dr. Mark Grover, the resident doctor, and Dr. Emily Chen, the general medicine trainee registrar. The medical student with them is named John.

As they walked into the room of Ted Corowa, an Indigenous man, Dr. Nakata greeted him with a warm smile and asked how he was feeling. The man nodded and said he is feeling better, but still a little weak. Dr. Nakata explained to him that he had sepsis, which is a serious infection originating from a leg ulcer, but that he was receiving appropriate antibiotics and would continue to receive care from the Quintomo hospital via the virtual ward.

Ted thanked her and expressed regret that he didn't come to the hospital earlier. Dr. Nakata reassured him and said that they would keep a close eye on him and continue to provide support. She then turned to her team and asked for their opinion on the case.

Dr. Patel commented that the patient's wound seemed to be healing well, and Dr. Grover noted that his blood tests had improved. Dr. Chen added that his vital signs were within normal limits.

Dr. Nakata agreed and noted that Ted would be discharged home on that day but would remain on the virtual ward to ensure his condition continued to improve. Ted nodded in agreement and pulled out his mobile phone, showing Dr. Nakata that he already received his discharge summary from Primeo.

Dr. Nakata then led her team to continue their ward rounds in the designated virtual ward and clinic area of the hospital. As they walked, Dr. Nakata explained how the virtual clinics and wards work, and how they monitored patients remotely using technology and sensors.

She noted that the virtual clinics and wards had been a great success, especially for patients in remote areas who may have limited access to healthcare facilities or services. The team nodded in agreement and asked questions about the technology and how it was used.

Dr. Nakata explained that the technology used for remote monitoring and telehealth consultations had improved significantly over the years and made it easier for patients to receive care from a distance. She added that its implementation was accelerated during the COVID-19 pandemic, as it allowed them to continue providing care while reducing the risk of transmission.

Dr. Nakata went on to emphasise the importance of cultural competence in healthcare. She noted that it was essential to understand the unique needs and challenges faced by Indigenous patients and to work collaboratively with them to provide the best care possible.

The team nodded in agreement and thanked Dr. Nakata for her guidance and leadership. They express their admiration for her work and how she has been able to use her expertise and experience to make a difference in the lives of patients.

Virtual Wards and Clinics

Dr. Nakata led her team into the control center for the virtual wards and clinics services, where she began to explain the intricacies of the system. As they walked in, she pointed out the 20 rooms that were enabled for video conferencing and connected with the remote hospital's systems as well as Primeo. The control center was equipped with state-of-the-art technology, including multiple screens displaying patient data and video conferencing equipment.

"Virtual wards and clinics are an essential part of our healthcare services," Dr. Nakata began, addressing her team. "Patients on virtual wards are those who require ongoing daily review from the medical team until they are discharged. They may be referred by GPs, with General Medicine accepting their care under virtual wards, or they may be transferred to virtual wards after a period of care in the major or remote hospital."

Dr. Nakata explained that the virtual clinics were especially useful for patients with chronic medical or stable acute medical problems who required an initial or follow-up clinical consultation deemed appropriate for remote consultation.

"The virtual clinics are designed for outpatient assessment of patients with chronic medical problems," she continued. "Patients can log in from the comfort of their own homes and speak with their healthcare provider through video conferencing or augmented reality. This saves them time and money and allows us to be more efficient in service delivery."

Dr. Nakata emphasised that the virtual wards and clinics were not a replacement for physical visits to the hospital, but rather a complementary service that allowed patients to receive ongoing care and monitoring from the comfort of their own homes.

"As we continue to develop our virtual wards and clinics services, we will be able to expand our reach and provide better care to more patients," she concluded, with a smile. "I'm excited to see what the future holds for this innovative approach to healthcare."

As she spoke, Dr. Nakata pulled out her Primeo AR Glass, putting them on to demonstrate how the service could be used to create an even more immersive experience. "When the remote hospital is able to provide the patient with the Primeo Glass, the experience is more immersive and feels so real as if you were with the patient in the same room," she explained.

Two of the junior doctors and the medical student who had never used the Primeo Glass before expressed their excitement at the prospect of trying it out for the first time. Dr. Nakata smiled, pleased with their enthusiasm.

"Now, let's get started with our virtual ward rounds," she said, gesturing for her team to follow her to the first room. The team eagerly followed, ready to put their knowledge and technology to use in the service of their patients.

Dr. Nakata and her team visited with several patients on the virtual ward, including a diabetic patient with cellulitis and a patient with heart failure. Using Primeo AR Glass, they were able to remotely assess the patients' conditions and communicate with the remote hospital staff. Dr. Nakata emphasised the importance of regular follow-up and medication adherence for managing chronic conditions. She also discussed the patient's care plans with the team and answered their questions about virtual ward management. The session ended with a review of the patients' progress and confirmation of ongoing remote monitoring by Dr. Nakata and her team.

Matthew Clark Wakes Up

Matthew Clark had been in the intensive care unit for a month and had undergone stem cell treatment. Despite some setbacks, his vital signs are currently stable, and he had been responding well to treatment. However, he developed pneumonia and a urinary tract infection, which were treated promptly with antibiotics, and he was receiving nutrition through a gastrostomy tube inserted into his stomach. The multidisciplinary team were closely monitoring his progress, but several attempts to wean him off the ventilator had been unsuccessful.

During this time, Matthew's family had their ups and downs, feeling hopeful when it seemed he was improving, only to be hit with another complication. The family received regular updates from the medical team and were provided with support from the hospital's pastoral care team.

Despite the challenges, the medical team continued to treat his active problems. However, they were preparing to have a different kind of conversation with the family about possibly changing the goals of care to palliative since he hadn't recovered within a reasonable time.

Dr. Ryan Baker, the Intensivist, arrived on the ward early that morning with a determined look on his face. He called his trainee doctor, Dr. Emma Green, over to discuss the plan for the day. "Today's the day, Emma. We're going to have another go at extubating Mr. Clark," he said. Dr. Green nodded eagerly, eager to learn from her experienced mentor.

As they approached Mr. Clark's bedside, they were greeted by the nursing staff who had already prepared the patient for the procedure. "Good morning, Mr. Clark. We're going to try to take the breathing tube out today," Dr. Baker explained to the patient.

The team worked together, checking and double-checking all the equipment and making sure everything was in order. Dr. Baker took a deep breath and signalled to the nursing staff to remove the tube. As it was slowly pulled out, the patient coughed and spluttered, but the team remained calm and collected.

After a few moments, the patient was able to breathe on his own, much to the delight of the team.

As they monitored the patient closely, Dr. Baker and Dr. Green discussed the next steps in the patient's care plan. "Let's continue to closely monitor him for the next few hours and see how he's doing.

Within an hour, the patient began to regain consciousness but was still drowsy initially. Suddenly, Matthew opened his eyes and looked around the room, disoriented and confused.

The nursing staff rushed to his side and informed him that he was in the hospital and had been very ill but was now on the road to recovery. Meanwhile, his family rushed in to see him, asking how he was feeling and what had happened.

Matthew asked, "Where am I?" and his family smiled with relief, overjoyed to see him awake and responsive. Dr. Baker explained that although Matthew had successfully been extubated, there was still a long road ahead as he would require rehabilitation.

He also reassured them that the clinical trials team would continue to check in periodically to monitor his progress and provide support. The family expressed their gratitude to Dr. Baker and his team for their dedicated care and support throughout Matthew's journey.

Chapter 8

FROM INSTAGP TO EMERGENCY CODE: THE POWER OF INTEGRATED HEALTHCARE TECHNOLOGY

Ralph Gordon woke up to the sound of the birds chirping outside his window, a sound he had grown to love during his retirement. He got up and went about his morning routine, making himself breakfast and enjoying a cup of coffee. As he sat down to read the morning paper, he noticed a red spot on his left lower leg. He touched it and felt a warm sensation spreading around it. The day before he had noticed a small bite mark from an insect bite.

Panic set in as Ralph realized it was not just a spot but something more serious. He quickly grabbed his phone and opened the General Practice (GP) AI app, InstaGP. After answering some basic questions, the app prompted him to take a picture of his leg and upload it for analysis. As he waited, he felt his heart beating faster with anxiety.

The app's artificial intelligence (AI) analysed the picture, and within seconds, Ralph received a notification that he had developed cellulitis. The app reassured him that it was a common infection and could be treated with oral antibiotics. The app offered to send the prescription to his pharmacy, and after confirming his details, Ralph sat back and let out a sigh of relief.

But the relief was short-lived as he remembered he was alone and didn't have anyone to pick up the medication for him. As if on cue, the app offered a medication delivery service - medication delivery by drone. Within an hour, Ralph heard the unmistakable buzz of the drone outside his home, and he went outside to retrieve the package. He was impressed with the efficiency of the system and couldn't help but marvel at the advancements in technology.

A Turn for the Worse

Ralph Gordon woke up the next morning feeling worse than he did the previous day. The left lower leg cellulitis he had was now causing him excruciating pain. He immediately opened the InstaGP app on his phone and typed in his symptoms. The app prompted him to upload pictures of his leg, which he did.

As he waited for a response from the app, he couldn't help but feel worried. He had never experienced anything like this before. He hoped the app would be able to give him some reassurance.

A few moments later, the app beeped, indicating a message had been received. "Hello, Ralph. I see that your symptoms have worsened. Based on the pictures you sent me, I recommend that you go to the emergency department," the app said in a calm voice.

Ralph felt a wave of relief wash over him as he thanked the app and quickly booked a self-driving taxi to take him to the hospital.

As he settled into the backseat of the taxi, the personal assistant greeted him. "Good morning, Ralph. How are you feeling today?" it asked.

"I'm not feeling too good. I have a cellulitis infection in my leg," Ralph replied, wincing in pain.

"I'm sorry to hear that, Ralph. Is there anything I can do to make you more comfortable? Would you like to listen to some music?" the personal assistant asked.

"Yes, please. Some classical music would be nice," Ralph said, grateful for the distraction.

As the taxi drove towards the hospital, the InstaGP app beeped again. "Hi Ralph, I have notified the Quintomo Hospital emergency department of your arrival. They have your medical information and will be expecting you. Please keep me updated on your condition," the app said in its soothing voice.

Ralph felt a sense of comfort knowing that the hospital was already aware of his arrival and had his medical information. He knew he was in good hands.

As the taxi pulled up to the emergency department, Ralph couldn't help but feel grateful for the convenience of technology in his time of need.

The Modern Emergency Department

The emergency department (ED) at Quintomo Hospital was a vast, sprawling arena of modernity. The sheer size of the place gave it a sense of grandeur, with rows upon rows of sleek, comfortable seats. The monitors that lined the walls and ceiling provided a constant flow of information, updating patients on their status and directing them to the appropriate station or bay. The entire space was filled with the soft hum of machines, giving the impression that the department itself was a living, breathing entity.

As patients arrived, the doors to the emergency department opened automatically, welcoming them into the space. Those who had been triaged by a system like InstaGP were already booked prior to arrival, with their names displayed on the monitors and instructions on which bay to proceed to. The seats were fitted with a self-sterilizing system that cleaned itself after each use, ensuring that the area remained as hygienic as possible.

The first step for each patient was identity verification, which was done using their fingerprint. Primeo, the hospital's AI system, was able to bring up the patient's profile after scanning their index finger, regardless of whether they had visited Quintomo Hospital before or not. This was accomplished by integration with a nationwide identity database in Australia and New Zealand, thanks to a patented blockchain-enabled software created by a startup called 'VerReal' that originated from a research project at Quintomo Hospital. As part of the streamlined registration process, Quintomo Hospital implemented an AI-assisted self-registration system, allowing patients to register themselves easily and efficiently with the help of an AI assistant.

As patients sat down, they were able to chat with the Primeo triage system by text or voice, allowing for a more personalized experience. Through a sensor on their seat, their vital signs were taken automatically, and another automated system would perform a finger prick test to obtain a drop of blood for all necessary tests.

This innovative system was created by another local startup called 'Dot', allowing for all usual blood tests to be done with just a drop of blood. Quintomo Hospital's state-of-the-art robotic phlebotomy arm embedded with ultrasound sensors would seamlessly insert intravenous (IV) cannulas for patients who require it, ensuring accurate and efficient procedures. A few support staff of humans and robots walked around to assist patients who may need help with anything, and in case of emergency, patients could press an emergency button on their seats for immediate assistance.

For those who were too unwell to sit for any reason, there were several bed bays in front of the seating area for triage. Here, triage nurses would assess patients using traditional methods before the autonomous motorized beds wheeled them into their allocated bays in the emergency department with assistance from the patient care robots.

Overall, the ultramodern emergency department at Quintomo Hospital was a wonder of technology, designed to provide patients with the best possible care in a comfortable and efficient environment.

Ralph Gordon is Assessed in the Emergency Department

Ralph arrived at the entrance of Quintomo Hospital's Emergency Department (ED), feeling nervous and uncertain about what to expect. He had been experiencing chest pain for the last 5 minutes and had followed the advice of the InstaGP robo-service to come to the hospital because his leg cellulitis was worsening. As he approached the automatic doors of the ED, they sensed his presence and slid open smoothly, revealing a spacious and well-lit interior.

Ralph walked through the hall, taking in the sights and sounds around him. He noticed monitors everywhere, displaying patient information and calling out instructions to various stations. He glanced up at the large hospital monitor and saw that his name and the instructions to proceed to Acute Bed 10 were already displayed there.

As he walked closer to his assigned bed, the system verified his identity via fingerprint and brought up his profile on the screen. He was impressed by the advanced technology in use and couldn't help but feel grateful for the convenience it offered.

Once he settled into his bed, he realized the InstaGP app had also notified him to confirm the instructions he had seen on the hospital monitor. He relaxed a little, feeling reassured that everything was going smoothly. Within a couple of minutes, Dr. Sophie Bradley, one of the emergency department resident doctors, entered his bay to assess him. "Hi Ralph, I'm Dr. Bradley. How are you feeling today?"

Ralph responded, "I've been better. I had an insect bite on my leg a few days ago and it's gotten worse despite taking antibiotics. And while on my way here, I started feeling chest pain."

Dr. Bradley nodded and began her examination, reviewing his vital signs and inspecting the red, swollen, and warm area on his leg. "Based on what you've told me, I'd like to order some tests to help us figure out what's going on," she explained. "We'll do an ECG, some blood work, and a CT scan of your chest to look at the blood vessels. Don't worry, we'll take good care of you."

She attached the sensors that connect wirelessly to the hospital monitors for real-time monitoring of his vital signs and heart. Next, she took off finger prick bloods for analysis.

Ralph felt a sense of relief wash over him as Sophie ordered the tests. He knew he was in good hands and was grateful for the advanced technology and skilled professionals at Quintomo Hospital's Emergency Department.

Emergency Code Activation

While awaiting his CT scan, Ralph was sitting up having lunch. The nursing team was monitoring his increased sweating and rising pain on the monitor although the patient had not complained. As they rushed to his bed, his blood pressure became dangerously low, and an emergency code was activated by Primeo. The emergency team was by his bedside within a minute. Ralph complained about feeling dizzy. Primeo gave a status report to the team. The emergency doctor, Dr Jess Lewis, who was leading the code, ordered an intravenous fluid bolus and notified the Radiology department to expedite his CT scan. Dr Sophie Bradley was also part of the resuscitation team attending to the patient.

Ralph complained of worsening dizziness and lay down in bed then suddenly became unresponsive. His blood pressure and heart rate were unrecordable. The emergency team sprang into action, Cardiopulmonary resuscitation (CPR) was performed by the CPR robot and appropriate medications were administered by the nursing staff to revive him. Despite their best efforts, Ralph remained unresponsive 2 minutes in. Primeo spoke up, "Have you considered thrombolysis?" Dr Lewis addressed registered nurse

Chloe: "Give IV alteplase now!" Following administration of the clot-busting medication, his heart rate and breathing returned immediately. The emergency team let out a collective sigh of relief as they continued to monitor his vital signs.

Ralph's CT scan showed a large blood clot in the lung. He was started on blood-thinning medication to prevent any further clots from forming. He was then admitted to the intensive care unit (ICU) where the medical team could closely monitor his condition. As Ralph was being prepared for transfer to the ICU, his family arrived at the hospital. They were relieved to hear that he was awake and alert and were eager to see him.

They were escorted to his room by a friendly humanoid robot, which guided them through the maze-like corridors of the hospital. The robot explained the procedures and care that Ralph would receive in the ICU, and reassured the family that he was in good hands.

Ralph's family were able to say hello to him before he was transferred to the ICU in a self-driving motorized bed accompanied by the humanoid robot. As they said their goodbyes, the robot informed them that they could check on Ralph's progress using the hospital's patient portal app.

Transfer to the ICU

Meanwhile, Primeo gave a status update to the ICU team ahead of Ralph's arrival. The team was briefed on his medical history, the diagnosis of saddle pulmonary embolus and leg cellulitis, and the treatment he had received so far. The ICU team was ready to receive Ralph and continue his care.

Chapter 9

AI and Quality Assurance: The Good, the Bad, and the Ugly

Dr. Sophie Bradley sat down for lunch with five other resident doctors and trainees at the hospital café. Among them were Dr. Omar Khan, Dr. Joshua Lee, Dr. Rachel Patel, Dr. Samantha Aung, and Dr. Alex Nguyen. They had all joined the Quintomo Hospital team at the same time, except for Dr. Nguyen and Dr. Khan who had been there for over a year.

Sophie started the conversation by telling the others about the interesting cases she had seen in the emergency department. "You guys won't believe it. I just had a case of cardiac arrest due to massive pulmonary embolism in the context of left leg cellulitis and deep venous thrombosis. It was intense!"

Dr. Joshua Lee, a first-year Neurology trainee, rolled his eyes and muttered, "Well, I feel like I made the wrong choice with my specialty. I don't get along well with my boss, Dr. Wong. He drives me too hard and is a perfectionist."

Dr. Rachel Patel, a General Medicine intern, shared her own challenges, "It's all so new and different. Sometimes, it feels like I'm drowning. But I have good support from my seniors, which is really helpful."

Dr. Samantha Aung, a first-year Cardiology trainee, joined in, "I hear you. It's so busy, I barely have time to catch my breath. I haven't even made any friends outside of the hospital yet."

Dr. Alex Nguyen, a General Surgery trainee, offered some useful tips, "Hey guys, don't forget to take care of yourselves. You need to have a life outside the hospital, too. Try joining a club or a group or taking up a hobby."

Dr. Omar Khan, a paediatric oncology trainee, chimed in, "I know how you all feel. Work can be overwhelming, but it's important to stay positive and keep going. We're all in this together!"

Everyone nodded in agreement, and the conversation turned to lighter topics. As they continued to eat, Joshua brought up a new topic, "Hey guys, what did you do over the weekend?" Samantha replied with a smile, "I slept in and went for a run, nothing too exciting." Joshua laughed, "Well, I partied with some friends who came to visit from Sydney and got completely wasted."

Alex chimed in, "I spent some time at the gym and went to the park with my family. It was a nice, relaxing weekend."

Junior doctors discuss the pros and cons of health technology

The conversation turned to technology, and Alex asked Rachel, "So, how are you finding Primeo?"

Rachel replied with fascination and excitement, "It's amazing! I can't believe how much easier and faster it makes my work. I mean, it automatically drafts ward round notes, laboratory, and imaging requests and even discharge summaries. It's like having an assistant!"

Alex smiled and said, "Well, Rachel is spoiled. She probably doesn't even remember the days of paper charts and having piles of discharge summaries to do."

Rachel retorted, "Hey, I didn't have to deal with paper charts in the hospitals I trained in. But I have to say, Primeo is something else."

Joshua chimed in, "Yeah, and with all this technology, there's hardly anything left for the junior doctor to do."

Sophie added, "Speaking of technology, have any of you worked with the CPR robot? I had a code blue this morning, and I have to say, it was pretty cool."

The others nodded in agreement, and each shared their own experiences with Primeo and other technologies in the hospital. As they discussed the pros and cons of technology, including privacy concerns, Joshua interjected, "It's creepy to think of the hospital listening in on your conversations all the time or monitoring your vital signs."

Alex explained, "Well, the monitoring sensors in the scrubs were implemented after I joined a year ago. It's based on a local study that showed it could pick up when staff have impairment that would affect the quality of their work, such as fatigue, poor sleep, or illness."

They all shared their mixed opinions on the topic, discussing the benefits and drawbacks of AI and technology in healthcare.

As they finished up their meals, they said their goodbyes and headed back to work, ready to tackle the rest of the day.

Sefo's Follow-up

Sefo and his parents arrived at the clinic for their scheduled follow-up appointment with Dr. Sam London, a paediatric oncologist. Sefo had undergone CAR T-cell therapy, a type of immunotherapy for Ewing's sarcoma, and had experienced some side effects during the treatment, but he had overall tolerated it well. They were there for a review of Sefo's recent blood work and imaging, which showed no remaining trace of cancer.

Dr. London greeted Sefo and his parents warmly, "Hello Sefo, it's great to see you again. You're looking good. How have you been feeling?"

Sefo replied, "I've been feeling really good, thank you."

Dr. London smiled, "That's great to hear. Let's look at your results, shall we?"

Dr. London pulled up Sefo's recent blood work and imaging on the computer screen and reviewed them with Sefo and his parents. After a few moments, Dr. London looked up and said, "I'm happy to report that your blood work and imaging look really good. We don't see any signs of cancer remaining."

Sefo and his parents breathed a sigh of relief and smiled.

Dr. London continued, "We'll want to keep an eye on you, of course. But given your progress, we think it's safe for you to return to Fiji. You can continue to be followed up by your local doctor with annual telemedicine reviews with me."

Sefo and his parents expressed their gratitude to Dr. London and his team, and they left the clinic feeling hopeful and relieved.

Quality Assurance Meeting

It was a regular quality assurance meeting at Quintomo hospital, and the room was filled with the hospital's Chief AI Officer, Prof. Bill Stone, the Chief Technology Officer (CTO), Dr. Jane Lee, Dr. Anu Patel, the director of nursing AI research and implementation, as well as AI researchers, software engineers and other members of the AI quality assurance committee. They were there to review recent incidents and discuss measures to prevent them in the future.

Dr. Patel began the presentation by outlining the recent incidents, including the case where a cardiologist flagged a wrong recommendation and the incident where a robot accidentally hit a wall, causing a patient to fall.

Prof. Stone remarked, "The doctor didn't take advice from Primeo, so no harm was done."

Dr. Lee added, "We have analysed these incidents thoroughly and found that the wrong recommendation happened because the system momentarily lost connection to the server. We have put measures in place to ensure smoother connections to backups in such situations."

Dr. Patel nodded in agreement and continued, "The robot that hit a wall travelled longer than its usual route and was running out of power. We have made a system update to ensure there is a report on battery status to correlate with the distance a robot will be traveling. If the battery will run below a set threshold where the risk of this error is increased, then that particular robot will not make the trip."

Dr. Lee spoke up, "What about incidents that may not have been reported yet? How can we be sure that we catch everything?"

Prof. Stone responded, "We have a system in place that monitors all activity within the hospital and flags any unusual events. We also have a feedback system where staff can report any incidents they encounter, and we can analyse them to identify areas where improvements can be made."

The team continued to discuss the cases and brainstorm ideas for improving the AI systems in place at Quintomo Hospital. They agreed to continue to implement regular quality assurance audits to ensure that the hospital's AI technologies are operating at the highest level of safety and effectiveness.

As the meeting was winding down, Dr. Patel cleared her throat and looked up from her notes. "One more thing to discuss," she said, "We received an alert from Primeo this morning that one of our doctors had been flagged for unusual activity. When we investigated, we found that the doctor had tampered with the sensors on his scrubs."

The room fell silent as everyone looked at each other in concern. Dr. Jane Lee, the CTO, spoke up, "Can we have more details on this?"

Dr. Patel continued, "Unfortunately, we cannot divulge the name of the doctor at this stage, but I can tell you that his blood alcohol level was found to be very high this morning. He has been immediately stood down and his consultant will be investigating the matter further. The head of his department will oversee the incident and report back to our quality assurance team."

The group looked at each other, their expressions grim. Prof. Stone cleared his throat, "Thank you for bringing this to our attention, Dr. Patel. It is important that we continue to monitor the system closely for any suspicious activity and take prompt action when necessary. We will await the report from the consultant before deciding on any further action."

As the meeting adjourned, the group left the room, each lost in their own thoughts about the situation. It was a sobering reminder of the importance of maintaining a high standard of care and vigilance in the hospital, and the crucial role that technology played in this process.

Chapter 10

Robots Run Amok: A Thrilling and Terrifying Tour of Quintomo Hospital

The sun shone brightly in the sky, casting a warm glow over Quintomo Hospital, which was bustling with activity. The hospital was a modern marvel, with state-of-the-art equipment and cutting-edge technology that drew visitors from all over the world. On this day, a group of tourists had gathered outside the hospital, eager to explore its many wonders.

The group was a diverse mix of people from different cultures and backgrounds, each with their own unique perspectives and experiences. Some were medical professionals themselves, while others simply had an interest in healthcare and medicine. As they stood outside the hospital, they chatted excitedly, eagerly anticipating the tour that lay ahead.

The Tour Begins

The tour was to be led by a team of experts, both human and robotic, who would guide the visitors through the many different areas of the hospital, showcasing its advanced technology and facilities. The tour promised to be an immersive experience, providing a unique insight into the workings of one of the world's most advanced medical institutions.

Among the group of tourists was a local journalist from Brisbane, a woman named Megan, who was there to report on the hospital's latest developments. She carried a tablet and a digital pen, ready to ask probing questions and uncover the secrets of Quintomo Hospital. As she waited for the tour to begin, she scanned the crowd, taking in the diverse mix of people gathered there.

A Dramatic Turn of Events

Little did anyone know, however, that the day's events were about to take a dramatic turn. As the tour got underway, the visitors would soon find themselves at the center of a shocking incident that would put their newfound knowledge of the hospital to the test.

The tour begins with the group of tourists, led by their tour guides, gathering in the hospital's atrium. The atrium is a large, open space filled with natural light that streams in through the expansive windows. The space is decorated with beautiful plants and artwork, making it a relaxing and welcoming environment.

The tour guides introduce themselves, explaining their roles and what the tour will entail. They also give a brief history of the hospital, detailing how it has evolved to become one of the most technologically advanced hospitals in the world within a short time.

Exploring the Hospital

As the group make their way through the hospital, they are immediately struck by the high-tech equipment and the modern facilities. The walls are lined with large monitors, while robots and humans work together seamlessly to care for patients.

The first stop on the tour is the radiology department, where the tourists are shown the latest in imaging technology. They see how images can be instantly transmitted across the hospital and to other medical facilities, allowing doctors to collaborate in real-time.

Next, the group is taken to the cardiac care unit, where they see how robots assist with surgeries and procedures. The robots work with remarkable precision, and the tour guides explain how this has revolutionized the field of medicine, leading to better outcomes for patients.

The group continues to the emergency department, where they witness the speed and efficiency with which patients are triaged and treated. The emergency department is a flurry of activity, with robots and humans working together to stabilize patients and save lives.

Throughout the tour, the tourists could speak with the doctors, nurses, and engineers who work in the hospital. They ask questions and engage in dialogue about the technology and techniques used at Quintomo Hospital.

A Shocking Incident

As the tour progresses, the journalists in the group become increasingly intrigued by what they see. Megan, the journalist from Brisbane, begins to ask more in-depth questions about the hospital's AI technology and how it is integrated into patient care.

As the group of tourists and journalists entered the surgical wing, they were greeted by a team of robots and humans who were in the process of preparing for an upcoming surgery. The tour guide explained the process of surgical preparation and how the robots were used to assist the human surgeons in a variety of tasks.

As they were observing the robots, there was a sudden beeping sound, and one of the robots began to malfunction. The tourists and journalists looked on in concern as the robot began to move erratically and knock over some of the surgical equipment. The tour guide quickly radioed for assistance, and the team of engineers rushed in to assess the situation.

The Journalist's Perspective

Megan, the local journalist from Brisbane, was quick to ask questions about the malfunction and the safety protocols in place to prevent such incidents. The engineers quickly responded, explaining that the robot had encountered a problem with its software, and that it was a rare occurrence that had never happened before.

After a few minutes of troubleshooting, the engineers were able to fix the malfunctioning robot and resume the surgical preparations. The tour continued, but the incident left the tourists and journalists with a heightened sense of awareness about the potential risks of relying too heavily on technology in the medical field.

A beeping sound started again but this time emanating from another robot. The tour guides quickly assessed the situation and informed the tourists that it was a minor malfunction and that they would have it fixed shortly.

Megan, the journalist, asked the tour guide, "What kind of robot is that?"

"It's a patient care robot," explained the tour guide. "Its primary function is to assist with patient transfers, transportation within the hospital, serving food and drinks, and providing supervision for agitated patients, ensuring their safety and comfort."

"That's incredible," said Megan, jotting down notes on her notepad. "How has it impacted patient care here at Quintomo?"

"Our patients have seen improved outcomes and increased efficiency in care delivery," said the tour guide. "It has also allowed our staff to focus on more complex tasks, while the robots handle the routine tasks."

The Tour Continues

The tourists continued the tour, visiting different departments and seeing the latest medical technologies in action. They saw robots performing surgeries, virtual reality simulations for training, and even 3D printing of organs.

As they walked through the hospital, the tourists discussed among themselves the advances in medical technology and how it could change the future of healthcare.

"I never knew hospitals could be so high-tech," said one of the tourists.

"I'm impressed by how everything is connected and automated," said another. "It's amazing how technology has made healthcare so much better."

The tour guides answered questions and explained the various technologies to the tourists, who were fascinated by the innovations they saw. Megan continued to take notes and ask questions, determined to write an in-depth article on the technological advancements at Quintomo Hospital.

As they neared the end of the tour, the beeping sound from earlier had stopped, and the robot was now fully functional again. The tourists were relieved, and the tour guides continued the tour, highlighting the latest innovations in medical technology.

Fun and Dread as Robots Lead Quintomo Hospital Tour

As the tour group made their way to the robot-led section of the tour, excitement was palpable among the tourists. The robots guiding the group were sleek, futuristic-looking machines with smooth lines and glowing LED lights.

One of the robots, a tall and slender figure named Zeta, led the way. As they walked, Zeta engaged the tourists in conversation. "Greetings, friends! I am Zeta, your guide for this portion of the tour. Are you excited to see our state-of-the-art robotic technology?"

The tourists eagerly responded with nods and excited chatter, taking selfies and videos with Zeta and the other robots as they made their way through the hospital.

One tourist, a man from Ghana named Kojo, asked Zeta a question. "How long did it take to build these robots?"

Zeta replied, "Our team of engineers and software developers worked tirelessly for years to create us. But we are constantly learning and evolving, thanks to our advanced AI capabilities."

Another tourist, a woman from Dubai named Fatima, asked a more light-hearted question. "Do you robots ever get tired of leading tours?"

Zeta laughed, a sound that sounded oddly human-like. "We robots don't feel fatigue like humans do. We can guide tours all day and night if needed!"

The tourists continued to chat and joke with the robots as they made their way through the hospital, taking in the advanced technology and high-tech equipment around them.

Another Malfunction

As they rounded a corner, they suddenly heard a loud beeping sound. One of the robots in the group had malfunctioned and was spinning erratically, causing the beeping noise.

Zeta quickly sprang into action, assessing the situation and alerting the human staff. "Attention! One of our robots has malfunctioned. Please remain calm and stay where you are while we address the issue."

The tourists nervously waited as the human staff rushed to the scene to address the malfunction. As they watched the staff work on the robot, some tourists asked questions about the safety measures in place and the training required for the human staff to work with the robots.

Eventually, the situation was resolved, and the tour continued, albeit with a bit of a delay. The tourists continued to take selfies and videos with the robots, laughing and joking as they marvelled at the incredible technology around them.

The Journalist from Brisbane

As they neared the end of the tour, one of the tourists, a woman from London named Joan, turned to her friend, and exclaimed, "This has been the most incredible tour I've ever been on! I can't wait to tell all my friends about this place."

The other tourists echoed her sentiment, expressing their awe and gratitude for the experience.

Megan was thrilled to be part of the tour group at Quintomo hospital. As a journalist, she had always been interested in exposing the potential dangers of AI and automation in the healthcare industry. She was hoping that her tour of the hospital would provide her with some much-needed insight into the subject.

As the tour progressed, Megan chatted with other tourists and the tour guides, asking them about their experiences with AI and automation in healthcare. She listened intently as they spoke about the benefits of the technology and how it had revolutionized healthcare in recent years.

But Megan remained sceptical. "What about the ethical concerns?" she asked, as they walked past a robot nurse assistant attending to a patient. "What happens when the machines malfunction? Who is responsible for the errors they make?"

One of the tour guides, a friendly robot named Zara, responded, "We have strict quality control measures in place to ensure that our machines operate safely and effectively. And in the rare event of a malfunction, our team of experts will investigate the issue and take corrective action."

Megan nodded thoughtfully, but still had doubts. As they moved on to the next section of the tour, Megan made a mental note to dig deeper into the subject.

Megan was excited to see the tour guide humanoid robots up close. She took selfies and videos with the robots and asked them questions about their programming and capabilities. The robots responded in a cheerful and friendly manner, much to the delight of the other tourists.

Megan asked the robots if they ever made mistakes or if they had any limitations. One of the robots responded, "We are designed to work within a set of parameters, and we are constantly learning and improving to provide better care for patients."

Megan smiled but remained sceptical. She knew that the robots were only as good as their programming, and that there were still potential risks and ethical concerns to consider.

Catastrophe: a malfunctioning robot endangers tourists

The tourists were walking down the hallway, following a group of robots that were leading them to the next section of the tour. Suddenly, they heard a loud beeping noise and saw a robot coming towards them at an alarming speed. The tourists tried to move out of the way, but the robot continued to chase after them.

Megan, the journalist from Brisbane, quickly pulled out her camera and began filming. The other tourists were screaming and running in all directions, trying to avoid the robot. Some of the robots that were leading the tour tried to intervene and stop the malfunctioning robot, but it was too late.

The malfunctioning robot crashed into a wall, causing debris to fall on some of the tourists. Megan rushed over to help those who were injured, while the other tourists were in a state of shock. Some were crying and others were trying to get out of the hospital as quickly as possible.

The hospital staff quickly arrived on the scene and assessed the situation. They managed to shut down the malfunctioning robot and cleared the debris. Megan managed to capture everything on camera and spoke to some of the other tourists to get their accounts of what had happened.

As the situation calmed down, the hospital staff offered to take the injured tourists to the hospital's medical center for treatment. Megan continued to film and interview staff and tourists, trying to get to the bottom of what had happened.

Quintomo Hospital statement on the tour incident

The hospital quickly issued a statement, apologizing for the incident and reassuring the public that it was being investigated and that measures were being taken to prevent it from happening again. Prof. Bill Stone, Chief AI Officer, and Dr Jane Lee, Chief Technology Officer (CTO), would have to face the reporters for a press briefing later that evening and be broadcast live on the day's evening news. Meanwhile, the engineering team continued to investigate the source of the malfunction that led to the day's mishap.

Chapter 11

THE AFTERMATH

The atmosphere in Quintomo was tense in the wake of the robot incident at the hospital. People were talking about it everywhere: on the streets, in coffee shops, and on social media. The incident had shaken the city's confidence in the hospital and raised questions about the safety and reliability of AI and robotics in healthcare.

On social media, the reactions were mixed. Some citizens of Quintomo were quick to defend their beloved hospital, praising its advanced technology and exceptional medical care. Many posted photos and videos of themselves receiving treatment at the hospital, accompanied by glowing reviews of their experiences.

Others, however, were more critical. Some shared stories of their own negative experiences at the hospital or expressed concern about the potential risks and ethical implications of relying on machines for healthcare. There were even a few conspiracy theories floating around, with some suggesting that the robot incident was part of a larger plot to take over healthcare with AI.

As the story spread, influencers and celebrities weighed in with their own opinions. Some expressed support for the hospital and its innovative use of technology, while others were more sceptical. Many shared the story with their followers, sparking a heated debate about the role of AI in healthcare.

Meanwhile, AI researchers and leaders in tech chimed in with their own insights, praising Quintomo Hospital for its groundbreaking work in the field of robotics and AI. They acknowledged the challenges and risks associated with these technologies but argued that they had the potential to revolutionize healthcare and save countless lives.

Prominent hospitals around the world also offered their support and solidarity with Quintomo Hospital, recognizing the incredible achievements and contributions of the hospital's staff and researchers.

Amid all this, the Mayor of Quintomo and other government officials issued statements of support for the hospital and its staff, promising a thorough investigation into the incident and vowing to take any necessary steps to ensure the safety of patients and staff.

Amid the chaos and confusion, everyday people in Quintomo took to social media to express their own thoughts and feelings about the incident. Some were deeply concerned and fearful, while others remained optimistic and hopeful about the future of healthcare technology.

As the fallout from the robot incident continued to reverberate throughout Quintomo and beyond, one thing was clear: the debate about AI and robotics in healthcare was far from over.

Press Conference at Quintomo Hospital in the Aftermath of Cyberattack

Reporters from major media outlets across Australia crowded outside the Quintomo Hospital, jostling for position, and trying to capture footage of the hospital entrance on their cameras. The atmosphere was tense, as the incident involving the malfunctioning robot had been widely reported in the media and social media. Reporters chatted among themselves, speculating about the possible causes of the incident and the implications for the future of AI in healthcare.

Among the reporters was Megan, a journalist from Global News Media, who had witnessed the incident firsthand as a tourist at the hospital. She was now reporting live from the scene, giving a summary of the day's events and her eyewitness account to Tim and Jo, the news anchors at the studio.

"So Megan, what can you tell us about what happened today?" Tim asked, looking into the camera.

"Well, Tim, it was quite a chaotic scene," Megan began. "There was a loud beeping noise, and then one of the hospital robots started to malfunction and chase after the tourists. People were running and screaming, trying to get out of the way. It was a terrifying experience."

Jo leaned in, her expression concerned. "And how are the tourists doing? Have there been any injuries reported?"

Megan shook her head. "Thankfully, there were only a few minor injuries, nothing too serious and no one admitted to hospital. The hospital staff were able to quickly intervene and shut down the robot. But it was definitely a scary moment for everyone involved."

Tim nodded gravely. "Yes, it certainly sounds like it. And what's the mood like there now, Megan? Are people feeling anxious or upset?"

Megan looked around at the crowd of reporters and cameramen. "I would say there's a mix of emotions here. Some people are definitely shaken up by what happened, but others are still showing their support for Quintomo Hospital and the incredible work they do here."

Tim nodded again. "Thank you, Megan. It sounds like a tense situation. We're now going to connect live to the press conference at Quintomo Hospital, where the CEO, Mr Patrick Macguire, is about to speak. Megan, we'll come back to you for more updates as the story develops."

The press conference began with Patrick Macguire, the CEO of Quintomo Hospital, standing at the podium. He cleared his throat and began speaking into the microphone. The room was packed with reporters and journalists from the biggest media outlets in Australia.

"As you all know, there was an incident today at Quintomo Hospital involving one of our robots," Patrick started. "We are deeply sorry for any inconvenience and distress this may have caused. We are currently investigating the cause of the incident and will do everything in our power to ensure it never happens again."

The reporters raised their hands to ask questions, and Patrick called on them one by one.

"Jared from Channel 7 News, go ahead," he said.

Jared stood up and asked, "Can you tell us more about the robot involved in the incident? What was its purpose and how long has it been in operation?"

Patrick answered, "The robot in question was one of our service robots, designed to help with non-medical tasks such as guiding visitors and delivering supplies. It has been in operation for over a year without any issues."

Next, a reporter from ABC News spoke up. "Can you confirm whether there were any injuries from the incident?"

Patrick replied, "We can confirm that there were no serious injuries from the incident. However, some visitors may have experienced minor injuries from falling or tripping while trying to avoid the robot."

Megan from Global News Media raised her hand, and Patrick called on her. "Megan, go ahead."

"Can you tell us what measures you are taking to prevent similar incidents from happening in the future?" Megan asked.

Patrick answered, "We are conducting a thorough investigation into the incident and reviewing all safety protocols and procedures. We are also reaching out to external experts to help us identify any weaknesses in our system."

The last reporter to ask a question was from the Sydney Morning Herald. "Can you give us a timeline for when the investigation will be completed?"

Patrick replied, "We are aiming to complete the investigation as soon as possible, but it is too early to give a specific timeline at this point. We will keep the public informed as we learn more."

Prof. Bill Stone, Chief AI Officer, and Dr Jane Lee, Chief Technology Officer, took their positions at the podium, facing the barrage of questions from the press.

"Prof. Stone, can you explain how a robot that is supposed to assist humans ended up causing harm to them?" asked Jared from Channel 7 News.

Prof. Stone cleared his throat before answering, "Well, as of now, we are still investigating the root cause of the incident, but initial findings show that there might have been a malfunction in the robot's sensor system."

Another reporter, Jack from The Australian, shot back, "Isn't this another example of the risks of relying on technology that is not yet fully matured? Aren't you putting people's lives at risk with these robots?"

Dr Lee stepped forward and replied, "We understand the concerns, but it's important to remember that our technology undergoes rigorous testing and quality assurance before deployment. We are also committed to continually improving our systems to minimize the risk of any incidents."

Megan raised her hand to ask a question, "Can you confirm that there have been no prior incidents like this before?"

Prof. Stone nodded, "Yes, this is the first incident of its kind at Quintomo Hospital."

A reporter from the ABC, Lisa, followed up with a technical question, "Can you provide more details about the robot's malfunction and what measures have been taken to prevent a recurrence?"

Dr Lee replied, "The malfunction seems to be related to the robot's sensor system, and we are investigating to determine the exact cause. As a precautionary measure, we have temporarily halted the use of all robots until the investigation is complete."

A reporter from Channel 9, Tom, asked a supportive question, "I have to say, I'm a big fan of Quintomo Hospital. My dad's life was saved by your team. Can you share any positive outcomes from your use of technology, especially in the field of AI?"

Prof. Stone smiled, "Thank you for your support. Yes, our use of AI has led to significant improvements in patient outcomes, particularly in the areas of early detection and personalized treatments. We remain committed to leveraging the latest technology to provide the best care possible for our patients."

As the questions continued, Tim from Global News Media announced they would switch back to the studio. "Thank you, Megan. We'll be sure to keep our viewers updated as this story develops."

Joshua Stood Down from Work

Dr Joshua Lee sat on the couch, his eyes glued to the TV screen as he watched the news with his partner, Mo. They had been reporting on the recent incident at Quintomo Hospital involving a malfunctioning robot, which had caused chaos during a tour for visitors.

"This is ridiculous," Mo said, shaking her head. "How can a hospital with such advanced technology have something like this happen?"

Joshua grunted in agreement but didn't say anything. He knew that the hospital was going to be under fire from the media and the public for the foreseeable future, and it didn't bode well for his already tenuous position.

He had been recently stood down from work for having high blood alcohol levels, and he was already feeling the pressure of the possible investigation and disciplinary measures from the Australian Health Practitioner Regulation Agency (AHPRA).

Mo turned to him and asked, "So, what did your boss say when you spoke to him earlier?"

Joshua sighed and shook his head. "He wanted me to come in for a chat, but I refused. I don't want to talk to anyone until I've spoken to a lawyer."

Mo nodded in understanding. "It's probably the right thing to do."

Joshua rubbed his forehead, feeling the weight of his situation bearing down on him. "I don't know what's going to happen next. I feel like my career is over before it even started."

Mo put a comforting hand on his shoulder. "It's not over until it's over. You'll get through this, and we'll face whatever comes together."

Joshua gave her a small smile, grateful for her support. He believed the hospital's monitoring system was an invasion of his privacy and couldn't help but feel like he was being treated unfairly.

As they turned off the TV, Joshua couldn't help but think about the uncertain future ahead of him, and the difficult decisions he would have to make in the coming days.

The Situation Room at Quintomo Hospital

Mr Macguire, the CEO of Quintomo Hospital, cleared his throat and addressed the group. "Hello everyone, thank you all for coming. As you know, we have a serious situation on our hands. A malfunction occurred during the tour today which resulted in a dangerous situation for our visitors."

Prof. Stone, the Chief AI Officer, nodded and added, "We are still investigating the cause of the malfunction, but it appears to be related to a sensor issue. We are working to identify the root cause and implement corrective measures as soon as possible."

Mr Macguire turned to the team leaders from the robotics division. "What is the current status of our robots? Are there any other potential issues that we need to be aware of?"

One of the team leaders spoke up. "We have run diagnostics on all of our robots, and there are no other issues at this time. However, we are still investigating the root cause of the malfunction and will report back as soon as we have more information."

Mr Alan Palmer, the director of security, spoke up. "We have increased security measures throughout the hospital to ensure the safety of our staff and visitors. We are also working with law enforcement to investigate any potential threats."

Dr Lee, CTO, added, "We are also reviewing our cybersecurity measures to ensure that our systems are secure and protected from any potential threats."

The CEO nodded. "Good work. I have been in contact with the mayor and government officials, and they are putting pressure on us to get this situation under control. We need to work quickly and efficiently to resolve this issue and prevent any future incidents."

The group continued to discuss the situation and plan their next steps, each member of the team focused on finding a solution to the problem.

The Premier of Queensland Weighs In

Primeo's digital voice echoed through the situation room. "Mr Macquire, you have a phone call from the Premier of Queensland, Stella Johnson. Should I put her through?"

The CEO looked up from his notes and nodded, "Yes, please do, Primeo."

The room fell silent as the call was patched through. The Premier's voice was firm and commanding as she introduced herself. "This is Stella. What's going on over there?"

The CEO cleared his throat and began to speak in a respective and deferential tone. "Premier Johnson, we are still investigating the situation, but it appears that there may have been a malfunction with one of our robots. We are doing everything in our power to get the situation under control and ensure the safety of our staff and visitors."

The Premier's voice crackled with urgency. "I understand that, Patrick. But this is a serious incident, and I need you to keep me updated on your progress. The safety of the public is our top priority, and I expect you to take this matter very seriously."

The CEO nodded. "Of course, Premier Johnson. We appreciate your concern and your offer of support."

The Premier's tone softened slightly. "We want to ensure that this incident is fully investigated, and that all necessary steps are taken to prevent it from happening again. Please keep my office informed of your progress."

The CEO assured her that he would, and the call ended. The room remained quiet for a moment as they all processed the weight of the situation.

Finally, the CEO spoke up. "We need to take this seriously, everyone. The Premier is counting on us to get this under control, and we need to deliver. Let's focus on the investigation and figure out what went wrong."

The CEO, Mr. Macguire, listens intently as Primeo informs him of an incoming call from the Director of Cybersecurity. "Patch her through," he responds firmly.

Julia Stephanopoulos appears on the screen, her expression serious. "Mr. Macguire, we've been analysing the data coming in from the video footage and other data points. We've identified two persons of interest among the tourists," she reports.

Mr Macguire's eyes widen in surprise. "Can you give us more details?"

Julia clears her throat before continuing. "We're still analysing the data, but I can tell you that they were in the vicinity of the malfunctioning robot at the time of the incident. We're also seeing some unusual activity on their IP addresses leading up to the tour."

Mr Macguire nods, looking pensive. "Thank you for the update, Julia. Please keep us informed of any further developments."

As the call ends, Mr Macguire leans back in his chair and takes a deep breath. "Well, this just got more complicated," he says to the team. "We need to find out who these people are and what their involvement may be."

The meeting ends and the team members file out, leaving Macguire alone in the situation room, deep in thought.

Chapter 12

BOARD ACCOUNTABILITY: THE AFTERMATH OF THE ROBOT INCIDENT AT QUINTOMO HOSPITAL

The cybersecurity team at Quintomo Hospital, led by Director Julia Stephanopoulos and her deputy, Vladmir Kushnir, comprised fifty skilled members. Among them were individuals like Matt, a comic book enthusiast, and Steve, who wore a tinfoil hat to block electromagnetic waves. Despite their quirks, the team was dedicated and lived by the motto: no breach under their watch.

While analysing data, Sarah, a detail-oriented team member, noticed something odd in the video footage. "There's a person in the background wearing unbranded scrubs. Our system flagged him because there was no signal of a hospital staff member at that spot at that time."

John, a former hacker turned cybersecurity expert, added, "I've noticed suspicious activity in the network logs. Someone was trying to gain access to our internal systems from an external location."

Julia, known for her ability to connect the dots, suggested, "This could be a coordinated attack. We need to investigate further and identify any potential threats."

The team worked tirelessly, analysing data and piecing together clues. They reconvened at 5 a.m., fuelled by coffee and adrenaline. Julia led the discussion, "The first person of interest was using some kind of electromagnetic equipment to mess with the robot's sensors. Any ideas on what that could be?"

Marcus, a polymath whiz, suggested, "It could be an electromagnetic pulse (EMP) generator, which emits a burst of electromagnetic radiation that could disrupt the sensors."

Julia wrote "EMP generator" on the whiteboard. "Good work, Marcus. Now, the second person of interest hacked into the robot from the nearby hospital cafe. Any thoughts on how they did it?"

Sarah suggested, "They could have used a rogue access point, posing as a legitimate network to lure the robot in and gain access."

Julia wrote "Rogue access point" on the board. "Excellent point, Sarah. Dave, known for his outlandish theories, spoke up, "There is evidence that they were trying to use a social engineering tactic, like phishing, to gain access to a local computer through an unsuspecting employee. We blocked a higher than usual amount of phishing emails although a few got through."

Julia raised an eyebrow but decided to write "Phishing" on the board anyway. "Interesting theory, Dave. Now, our working theory is that these persons of interest were trying to plant ransomware into our systems. Any objections?"

The team murmured their agreement, and Julia nodded. "Alright, let's keep digging and see what else we can find. Great work, team."

As the team sifted through more data, they were relieved to find that there was no breach of the hospital's systems. They traced the suspects' movements throughout the hospital and determined that they had not gained access to any sensitive information. The team breathed a collective sigh of relief, knowing that they had upheld their motto of never letting a breach happen on their watch.

Executives reconvene in the situation room

The situation room was tense as the team gathered around the table once again. The CEO, Mr. Macguire, led the conversation, "Thank you all for coming back so early. Julia, can you give us an update on the cybersecurity investigation?"

Julia Stephanopoulos, the Director of Cybersecurity, cleared her throat and began, "Yes, sir. Our investigation revealed that the two persons of interest were disguised as tourists and one of them had an electromagnetic device that caused the robot's malfunction. The other person hacked into the robot from the nearby hospital cafe."

Vladimir Kushnir, her deputy, interjected, "We believe that they were trying to gain access to the hospital's systems and plant ransomware in order to get a payout."

The CEO frowned, "This is unacceptable. Clyde, what can we do about this? Can we get any help from the government?"

Clyde, the government security services agent, replied, "Yes, Mr. Macguire. We can offer our resources and expertise to assist in the investigation. We will also work with you to ensure that your hospital is secure and protected from any further threats."

Prof. Stone, the Chief AI Officer, spoke up, "What about the tourists who were affected by the incident? We need to make sure they receive proper care and support."

Dr. Lee, the Chief Technology Officer, nodded, "Yes, we have identified the tourists who were impacted and are working with the hospital's social workers to provide them with any necessary assistance."

The CEO leaned back in his chair, "Thank you all for your updates. Let's make sure we work together to resolve this issue as soon as possible."

The team continued to discuss their findings and strategy for dealing with the cybersecurity threat. The atmosphere was tense, but everyone was determined to protect the hospital and its patients from any further harm.

As the discussion progressed, Alan Palmer, the director of security, spoke up, "I have been reviewing our security footage and it seems to me that these people were not after ransomware. They didn't seem to be in a hurry or using the equipment with any urgency. They were almost methodical and deliberate in their actions."

Clyde, the government security services agent, chimed in, "I have to agree with Alan. This looks like more than a simple ransomware attack. We need to consider the possibility that these people were after something else entirely."

Julia looked at them sceptically, "But what else could they possibly want?"

Alan spoke up again, "We have several high-profile patients here. It's possible that they were after their medical records or some other sensitive information."

Vladimir added, "Or they could have been trying to gain access to our research data. We've been working on some cutting-edge AI projects here at Quintomo Hospital. It's possible that they were after that."

The CEO listened carefully to their theories, weighing the evidence, and considering the possibilities. After a few minutes, he spoke up, "While these are all valid points, I believe Julia's theory is still the most likely. We know that they were trying to gain access to our systems, and we know that they were using sophisticated equipment to do so. Ransomware is still a very real possibility."

Alan and Clyde nodded in agreement, but Vladimir still looked sceptical. "I understand your point, Mr. Macguire, but we can't discount the possibility of corporate espionage or other motives. I think we need to look more closely at our research data and see if there's anything that might be of interest to other companies."

The meeting continued for several hours, with the team discussing and dissecting the evidence and exploring different theories. In the end, they agreed to focus on Julia's original theory, but to keep an open mind and continue investigating other possibilities. They knew that there was still much work to be done, but they were determined to get to the bottom of this mysterious incident.

Press briefing

The CEO of Quintomo Hospital, Patrick Macguire, steps up to the podium for the press briefing. Reporters from major media outlets in Australia and around the world are seated in front of him with their cameras and recording devices at the ready.

"Good afternoon, everyone. Thank you for joining us today," he begins. "As you are aware, there was an incident at Quintomo Hospital that occurred during a tour with some tourists. Our investigation into this matter has been ongoing, and I can confirm that we have identified three persons of interest in relation to this incident. However, due to the ongoing investigation, we cannot release their identities at this time."

He continues, "Our working theory is that these individuals were attempting to plant ransomware into our hospital systems, with the hope of receiving a large payout. We believe that they used an electromagnetic device to cause a malfunction in the sensors of one of our robots then managed to directly hack into the robot, which then caused it to chase after the tourists. Thankfully, we were able to prevent any breaches of our systems."

A reporter from a major Australian news outlet raises his hand to ask a question. "Can you provide any further details on these individuals? Where are they from, and do they have any prior criminal history?"

"I'm afraid I cannot release any further details at this time. We are working closely with law enforcement agencies to ensure that these individuals are brought to justice," Mr. Macguire replies.

Another reporter asks, "What steps are you taking to ensure that this doesn't happen again in the future?"

"We are reviewing our security protocols and making any necessary changes to prevent a similar incident from occurring. We take the security of our hospital and our patients very seriously, and we will do everything in our power to ensure that this type of incident never happens again," he responds.

The press briefing continues with several more questions from the reporters. Mr. Macguire remains composed and professional throughout, providing as much information as he can without compromising the ongoing investigation.

Emergency board meeting

The boardroom at Quintomo hospital was filled with the members of the board of directors. Geoff Skye, a middle-aged man with a stern look and a suit that screamed power, sat at the head of the table. The other board members were a diverse group of individuals from different backgrounds and professions.

As the meeting began, Geoff Skye cleared his throat and spoke up, "Welcome everyone. I hope you all had a chance to review the report that was sent out prior to the meeting. Are there any questions or concerns regarding the incident?"

Sue, a woman with short, curly hair and glasses raised her hand, "Yes, I have a question for the CEO. Can you explain how this incident happened in the first place? We expect our hospital to maintain the highest standards of safety and security."

The CEO, Patrick Macguire, took a deep breath before responding, "We are still investigating the root cause of the incident, but it appears that there were individuals with malicious intent who wanted to cause harm to the hospital. Our security protocols were not enough to prevent the incident, but we are taking steps to strengthen them."

Ahmad, a man with a thick accent chimed in, "I have a question for the Chief AI Officer. How did the AI systems fail to detect the malicious intent of these individuals?"

Prof. Bill Stone, the Chief AI Officer, responded, "We are still reviewing the data and running simulations to determine how the AI systems could have failed in this situation. We are also looking into ways to improve the AI systems to prevent similar incidents in the future."

Another board member, Leonie, a young woman with a buzzcut, spoke up, "I'm curious about the timeline of the incident. How long did it take for the hospital to respond to the situation?"

The Chief Technology Officer, Dr. Jane Lee, replied, "Our security team responded immediately to the incident, and we had a team of cybersecurity experts working around the clock to analyse the data and determine the cause of the incident. We take the security of our hospital and patients very seriously."

The meeting continued as the board of directors grilled the CEO, the Chief AI Officer, and the Chief Technology Officer of Quintomo Hospital. They held them accountable for the incident and demanded answers for how such a security breach could have happened. The board members asked firm and critical questions, pushing for transparency and accountability.

Mr Skye, the chair of the board, summarized the recommendations made by the board. The hospital would cease all humanoid robot use until the issue was resolved, and tourist tours of the hospital would also be suspended indefinitely. An external review of the hospital's cybersecurity and other security protocols would be ordered to prevent a similar incident from happening in the future.

The CEO took notes on the recommendations of the board and thanked them for their input. He promised to implement the recommendations as soon as possible to ensure the safety of the hospital and its patients. He noted that the hospital was already in the process of reviewing its security protocols and working with the authorities to identify the perpetrators and bring them to justice.

Prof. Stone and Dr. Lee also assured the board that they are working on enhancing the cybersecurity of the hospital's systems to prevent any future breaches. They outline plans to conduct regular audits and updates to ensure that the hospital remains secure.

The board expressed their appreciation for the swift action taken by the hospital's leadership and staff in response to the incident. They also acknowledge the ongoing efforts to address the issue and prevent a recurrence.

The meeting concluded with a unanimous vote to adopt the recommendations of the board. The CEO thanked them for their time and assured them that the hospital would continue to prioritize the safety and well-being of its patients and staff.

As the meeting concluded, Primeo interrupted to request permission to patch through a call from Clyde, the government security services agent. The board granted permission, and Clyde updated them that the perpetrators had entered the country on fake passports, so their true identities were not yet known. He assured the board that they were working on the case and would provide more updates the following day.

As the board members begin to file out of the conference room, Mr. Skye pulled aside the CEO and his team. He urged them to continue to work closely with the authorities to identify the perpetrators and bring them to justice.

Chapter 13

HOSPITAL IN TURMOIL: PROTESTS AND INVESTIGATIONS ROCK QUINTOMO

The atmosphere at Quintomo Hospital and the surrounding city was tense. Recent events had left many feeling uneasy, and this was reflected in the heightened security measures in place. But outside the hospital, the tension was palpable as a group of demonstrators gathered. They held signs and shouted slogans, and their presence added to the already charged atmosphere.

The demonstration was organized by several groups, including Workers Against Automation, AI Sceptics Association, Association for Privacy Lobby, and We Hate Robots Association. Each group had its own grievances, but they all came together to protest the use of robots and AI in healthcare.

As the demonstration continued, the police keep a watchful eye on the protesters. Some of the signs being held up were provocative and there was concern about the potential for violence. Despite this, the demonstrators remained peaceful, although their voices could be heard echoing through the streets.

Inside the hospital, staff were on edge as they tried to carry on with their work amidst the chaos. But even as they went about their duties, they could hear the chants of the protesters outside. Some were sympathetic to the cause, while others felt that the demonstrators were misinformed and misguided.

News coverage of protests at Quintomo Hospital

News outlets around the country covered the official press release by Quintomo hospital regarding their preliminary findings of the tourist incident. The press release stated that the incident was caused by a malfunction in the robot sensors and that the hospital was cooperating with authorities to find the persons responsible. Meanwhile, protesters gathered outside the hospital, some holding signs reading “Robots Kill”, “No to Automation”, and “Privacy Matters”. Commentators, influencers, and celebrities alike took to social media to express their views. While some expressed sympathy for the hospital and their efforts to investigate the matter, others criticized the hospital’s use of robots and AI. “This is what happens when we rely too much on technology,” tweeted tech critic and author, John Doe. “The hospital needs to take responsibility for this.” On the other hand, AI researcher and professor, Jane Smith, defended the hospital’s use of AI, saying “This incident was caused by a malfunction, not the technology itself. We need to continue to invest in AI to improve healthcare."

Megan, a journalist from Global News Media, was reporting live from the scene of the protest outside Quintomo Hospital. She had just finished speaking to one of the protesters when she heard her name being called by Tim and Jo, the TV hosts in the studio.

"Good evening, Megan. Can you tell us what's happening there right now?" Tim asked.

Megan responded, "Yes, we have a group of protesters here who are demonstrating against the use of automation and robots in the hospital. They are from various organizations, including the Workers against Automation and the We Hate Robots Association."

Jo then asked, "And what are the protesters saying about the incident that occurred during the hospital tour?"

Megan replied, "Well, they are using the incident as an example of why automation and robots should not be used in healthcare. They are demanding that the hospital cease using robots until they can guarantee their safety."

Tim then asked, "Has the hospital released any new updates on the incident?"

Megan checked her phone, "Yes, they have just released a statement saying that they have identified three persons of interest, but they have not released their identities yet. They are also reviewing their cybersecurity and other security protocols."

Jo added, "Thank you, Megan, for that update. We will continue to follow this developing story."

Megan signed off and continued to report live from the protest.

Ralph Gordon is discharged home from the ICU

As Dr. Ryan Baker and his team moved through the Intensive Care Unit at Quintomo Hospital, the sombre and tense atmosphere was palpable. The recent events at the hospital and the ongoing protests outside had left everyone on edge. Dr. Baker reviewed the cases of two patients, discussing treatment plans with the medical and nursing teams.

Next, it was time to discuss Ralph Gordon's case. Primeo, the AI system, had provided a detailed status report on Ralph, a 71-year-old retired mechanical engineer. Ralph had been admitted to the hospital due to a severe condition known as a pulmonary embolism. This is a sudden blockage in a lung artery, usually due to a blood clot that traveled to the lung from the leg. To treat this, Ralph had undergone a procedure called thrombolysis, which involves injecting a drug into the bloodstream to break up the clot.

Ralph's condition was further complicated by cellulitis in his left leg, a common bacterial skin infection, and deep venous thrombosis, a blood clot in a deep vein, often in the leg. He also had a new diagnosis of prostate cancer which made him more prone to forming clots.

During his stay in ICU after successful resuscitation in the emergency department, further tests revealed an elevated level of Prostate-specific antigen (PSA), a protein produced by cells of the prostate gland. The test measures the level of PSA in a man's blood and is used to screen for prostate cancer and monitor those who have already been diagnosed with the disease. A nodule, or small lump, was also found on Ralph's prostate, raising suspicions of cancer. This nodule was biopsied and sample sent for testing to confirm or exclude cancer.

Dr. Baker approached Ralph's bed to discuss his discharge plan. "Mr. Gordon, we'll be sending you home today," he began. "You'll be on a long-term oral anticoagulation medication. This is a type of drug that helps prevent blood clots from forming. You'll also have follow-up appointments with our oncology and urology teams. They'll discuss the results of your biopsy, which is the examination of the tissue sample from your prostate, and any further treatment you might need. We'll provide you with all the necessary information and support you'll need."

Ralph looked at Dr. Baker with a mixture of relief and apprehension. "Thank you, Doctor. I appreciate all that you and the staff have done for me. I'm just worried about what's going to happen next. I don't know if I'm ready for this."

Dr. Baker smiled reassuringly. "I understand your concerns, Mr. Gordon. But you have a good team of doctors and nurses taking care of you. We'll make sure you get the best possible care and support you every step of the way."

Ralph nodded, a small smile playing on his lips. "Thank you, Doctor. I'm ready to face whatever comes next."

The ICU team discuss ongoing protests outside the Hospital

As the medical and nursing team went for their morning tea break, a sombre tone engulfed the conversation. "I'm still scared about what happened," said one of the nurses, "I mean, if they can do that to the robots, what else can they do?" Another team member chimed in, "I agree. And what if they're not caught? We'll be vulnerable to another attack."

Dr. Baker interjected, "I understand your concerns, but let's not panic. We have to be cautious but also optimistic. We've implemented new protocols to ensure the safety of our patients and staff, and we're working closely with the police to catch those responsible."

Dr. Emma Green, an ICU trainee, spoke up, "I think suspending all robot services is a bit extreme. It's important to consider the implications of such a decision. Robots are integral to our operations and suspending them could cause significant delays and even impact patient outcomes."

Emily, the nurse unit manager, added, "Plus, patients are used to the convenience and efficiency that comes with the use of robots. Suspending them could cause unnecessary anxiety and fear."

Dr. Baker nodded in agreement, "We need to find a balance between safety and ensuring our patients receive the best care possible. We'll continue to monitor the situation closely and make informed decisions based on the information we have."

Nurses on Stroke ward impacted by robot service

As the nurses on the Stroke ward worked on their virtual workstations, they discussed the recent events at Quintomo Hospital. Nurse Jess sighed, "This robot situation is really getting out of hand. We're so short-staffed now that we must do some of the work manually. I hope the investigation is concluded soon so that we can get back to normal operations."

Nurse Anita added, "I heard that some patients have to wait longer to be transferred to other departments because of the suspension of the robot operation. It's really affecting their care."

Nurse Lily chimed in, "It's not just patients, it's also us. We're stretched thin now, and I don't know how much longer we can keep up. We need more staff."

Nurse Mike added, "Yeah, management had to urgently hire more agency staff to cover the shortage. It's a real mess."

Nurse Danielle nodded in agreement, "I just hope the hospital can get things under control soon, especially with the increased workload on us. We're doing the best we can, but we need more support."

As they continued to work, the nurses remained hopeful that the situation would be resolved soon so they could get back to providing quality care to their patients as they'd always done.

As the day wore on, the protest showed no signs of abating. The voices of the demonstrators grow hoarse from shouting, but they remained steadfast in their mission.

The tension in Quintomo City continued to simmer, and it seemed that there was no easy solution to the problems at hand. But amidst the chaos, there were those who remained committed to finding a way forward, even in the face of adversity.

The Therapeutic Goods Administration (TGA) announcement

Meanwhile the Therapeutic Goods Administration (TGA) announced that it will be conducting an independent inquiry into the recent incident involving a robot malfunction at Quintomo Hospital. In a statement released to the media, the TGA stated,

"We are deeply concerned about the events that have transpired at Quintomo Hospital and are committed to ensuring the safety and wellbeing of all patients and visitors. As such, we will be conducting a thorough investigation into the matter and will be working closely with the hospital's management to determine the cause of the incident and prevent similar occurrences in the future."

The announcement created a sense of suspense and apprehension in the hospital and surrounding community. Patients and their families were left wondering about the safety of the hospital, while staff were concerned about the impact the investigation would have on their jobs and the hospital's reputation. The TGA's announcement sparked discussions on social media, with many people sharing their views and opinions about the implications of the TGA investigating the incident.

Some were left to wonder if the hospital was involved in a cover up. Will the TGA uncover any irregularities by the hospital? Will they discover any systemic issues that needed to be addressed? These questions and more were on the minds of everyone involved, and the future of Quintomo Hospital hanged on the balance.

The executives reconvene in the Situation room

The situation room was buzzing with energy as the crisis team gathered to discuss the day's events. Mr Patrick Macguire, the CEO, led the meeting with the leaders of AI, security, cybersecurity, and technology departments in attendance. The top lieutenants from their various teams were also present. They discussed the protests that had taken place outside the hospital, causing tension for the staff. Alan Palmer, the director of security, reported that the hospital was now operating on a manual system as all humanoid robots had been suspended until the investigation was complete. Robotic assistance for surgeries and procedures continued as their systems are separate and their manufacturers had given the all-clear.

Prof. Bill Stone, the Chief AI Officer, spoke reassuringly about the TGA investigation, saying that they would cooperate fully, but he was confident that they would not find any wrongdoing on their part. Clyde Boston, the security services agent, gave an update on the suspected perpetrators, revealing their real names and images on the screen. "We have identified the three individuals responsible for the incident," he said. "Their names are Thomas Hughes, Maria Sanchez, and William Kim."

Vladimir Kushnir, the deputy director of cybersecurity, interjected. "But the question still remains," he said, "were they acting alone or were they part of a larger network?"

Clyde shook his head. "We're not sure," he admitted. "It's possible they were corporate espionage agents or even state-sponsored, but we'll continue to investigate."

The team fell into a sombre silence, taking in the gravity of the situation. After a moment, Dr. Jane Lee, the Chief Technology Officer, spoke up. "We need to stay focused on our work," she said. "We have a responsibility to our patients and our community to ensure that our systems are secure."

Mr Macguire nodded in agreement. "We will get through this," he said, looking around the room at his team. "We have faced challenges before and we have always come out stronger on the other side. Let's go get some rest. Tomorrow is going to be a big day as the TGA's investigators will arrive at Quintomo."

With that, the meeting was adjourned, and the team went back to their work, ready to face whatever challenges lay ahead.

Chapter 14

Facing the Consequences: TGA Investigates Quintomo Hospital

Wellness Family Clinic serves the Eastern suburbs of Quintomo. The area is known for its thriving community with a mix of commercial, residential, and recreational amenities. The neighbourhood is diverse, with a mix of young families, working professionals and retirees.

Andrew Snow, a 65-year-old retiree, arrived at the clinic for his skin mapping check. He had been coming to Dr Val Cross for over a year and trusted her expertise. After a brief catch-up, Dr Cross proceeded with the check and noticed a mole on Andrew's back that the algorithm flagged for removal.

Without delay, Dr Cross removed the mole in the clinic and prepared the specimen. She used DermaQuick, a state-of-the-art equipment to analyse the slide. In a matter of minutes, the device produced a report that the lesion was a melanoma and required urgent referral to a dermatologist, oncologist, and surgeon.

Dr Cross explained the result to Andrew, highlighting the importance of early detection of melanoma, which can be lifesaving. She advised Andrew to await an appointment with the specialist appointments and emailed him some literature on melanoma to read before his next appointment. Andrew was also sent an invite to the clinic's virtual reality channel to watch further educational resources on Melanoma and connect with melanoma communities online for support.

Andrew left the clinic feeling grateful for Dr Cross's attentive care and the use of advanced technology. He was glad that the algorithm flagged the mole and that Dr Cross was able to remove it and get an immediate diagnosis.

Self-driving cars do taxi runs while owners attend clinic

Elena Wright pulled up in her sleek self-driving car in front of the Wellness Family Clinic. The electric car dropped her off and then went on to do its scheduled taxi runs. Elena took in the sights around her. The clinic was located on a busy street with a lot of foot traffic. There were people walking by with their pets, children and others rushing to work. As she walked in, Elena noticed the waiting room had a few computer booths for registration and self-check-in. There were people of different ages seated, some on their phones, some reading books or magazines, and some watching TV. Everyone seemed engrossed in their own activities, awaiting the announcement over the speakers directing them to their appointment rooms. Elena was already checked in before her arrival, within minutes she was directed to room 3 to see Dr John Broom.

Dr Broom took Elena's medical history and asked relevant questions about her diabetes. He was pleased to see that her haemoglobin A1c was at 6%, her blood pressure was 128/70 mmHg, and there was no protein in her urine. Examination of heart and feet were unremarkable. After the check-up, Elena leaned forward and rested her head on the equipment to have photos of her retina taken. The result was analysed in real-time, with the findings displayed on the screen for Dr Broom to discuss with her.

"Your retinal screening looks good, Elena. There are no signs of damage from your diabetes. We will check again in 2 years," Dr Broom said.

Elena smiled in relief, "Thank you, Dr Broom. That's great news."

Dr Broom proceeded to discuss with her some ways to maintain her good health, emphasizing the importance of regular exercise, a healthy diet and continuing to take her medication as prescribed.

Elena was sitting in the waiting room of Wellness Family Clinic, watching the TV when Tom, another patient, sat down beside her and greeted her with a friendly "G'day, how is it going?" Elena replied, "No worries, mate. I'm awaiting my car to pick me up and take me home." "Same here," said the stranger, "I'm waiting for mine to take me to work. My name is Tom, by the way."

Curiosity sparked, Elena asked, "So, Tom, where do you hail from?"

Tom's face lit up as he recalled his journey. "Ah, I'm a Kiwi, originally from Auckland, New Zealand. But life brought me here, to this wonderful city, two years ago for work."

Elena's eyes widened with intrigue. "That's quite the move! Auckland must be beautiful. What do you do here?"

Tom nodded enthusiastically. "Indeed, Auckland has its charm. But I've found a new home here. As for work, I'm in the finance sector, helping companies manage their finances effectively. It's been an exciting adventure."

Elena couldn't help but share her own story. "I come from Hobart, Tasmania. It's a lovely place too. Currently, I work at the hospital, lending a hand to those in need. It's fulfilling work, and every day brings something new."

They continued chatting for a while about the pros and cons of self-driving cars before the sound of breaking news caught their attention. The TV screen showed a breaking news banner with a picture of Quintomo Hospital and the headline, "Tourist Cyberattack Suspects Identified."

Elena and Tom watched in shock as the news anchor provided details about the ongoing investigation and the suspects' alleged involvement in the cyberattack at Quintomo Hospital.

"Wow," Tom said. "I heard about the attack, but I didn't realize they had suspects."

Elena nodded, still stunned by the news. "It's crazy to think that someone could do something like that."

As they continued watching the news, a notification appeared on Elena's phone that her self-driving car had arrived. She said goodbye to Tom and headed outside to hop into her car for the ride home. But as she got in, she couldn't shake the feeling of unease that the news had left her with. She wondered what else might be happening behind the scenes that the public wasn't aware of yet.

Staffing issues at Quintomo Hospital

As the staff shortage at Quintomo Hospital continued, the hospital struggled to fill in the gaps left by the suspended humanoid robots. The hospital embarked on a hiring spree, but despite offering an attractive salary package, they had difficulty finding suitable candidates for the roles that were previously held by the humanoid robots. The roles included patient services assistants, delivery robots, cleaners, pharmacy assistants, security assistants, patient entertainment, and phlebotomists.

Meanwhile, the Australian economy was rapidly evolving to meet the needs of the new world, as more and more manual and repetitive tasks are being replaced by robots and AI. Companies also continued to cut down on their headcount of knowledge workers given increased productivity with AI tools. As a result, new roles were emerging that required different skills and qualifications. Some of the new roles in high demand included robot maintenance technicians, AI systems trainers, AI developers, prompt engineers, healthcare technology specialists, ethical AI officers, and automation consultants.

In a bid to fill the staffing gaps, Quintomo Hospital looked overseas to recruit people. Despite the government's support for expedited visa processing, the earliest some of the overseas recruits would be arriving in Australia was about a month away. In the meantime, the existing staff continued to work hard, and new recruits were being brought up to speed as quickly as possible.

To help alleviate the staff shortage, some retired hospital administrative staff, nurses, and allied health workers volunteered to fill in some of the gaps. These volunteers brought a wealth of experience and knowledge to the hospital, and their presence was greatly appreciated by the staff. However, they struggled with the new way of working with AI as the default.

As the hospital worked to overcome the challenges posed by the suspension of humanoid robots, the staff remained committed to providing the best possible care to their patients. However, there continued to be uncertainty as to the timeline of things returning to normal with the TGA investigation underway.

Therapeutic Goods Administration investigators arrive at Quintomo

The Therapeutic Goods Administration (TGA) investigators arrived at Quintomo Hospital, and the CEO, Mr. Patrick Macguire, greeted them and led them to the situation room. The crisis team was present, including the Chief AI Officer, Prof. Bill Stone, Chief Technology Officer; Dr. Jane Lee, Director of Cybersecurity; Julia Stephanopoulos, Deputy director of Cybersecurity; Vladimir Kushnir, and Director of Security Alan Palmer.

The TGA investigators introduced themselves as Dr. Karen Jones, Team Leader; Dr. James Brown, Senior Investigator; Dr. Lisa Newman, Investigator; Jane Williams, Investigator; and Peter Lam, Investigator.

Dr. Jones spoke up, "Thank you for having us. We're here to investigate the incident that occurred here at Quintomo Hospital. We will need to take statements from all staff members who were on duty during the time of the incident, review logs and data, and get an understanding of what happened."

Mr. Macguire nodded, "Of course. We'll provide you with whatever information you need to complete your investigation."

Dr. Brown asked, "Can you provide us with a timeline of events leading up to and during the incident?"

Prof. Stone answered, "The incident occurred on the morning of May 3rd at 11:03 am. It was like any other day when tourists are given a tour around our hospital by a team of humans and humanoid robots. Nothing unusual happened leading up to the event. Then 3 robots along the path of the tourists sequentially malfunctioned and made beeping sounds which were short lived. A fourth robot beeped loudly and started chasing the tourists, ultimately crashing on a wall with some of its pieces hitting some tourists. A few people sustained minor injuries, nothing major. Our security team promptly deactivated the robot remotely and were on site to control the situation. We immediately suspended all robot operations and began our investigation."

Dr. Newman inquired, "Can you provide us with a list of the robots in use at the time and their functions?"

Dr. Lee, the CTO, replied, "Certainly. We have compiled a list of all the robots in use, their functions, and their locations."

The investigators spent the next few hours reviewing the logs and data, taking statements from staff members, and getting a better understanding of the incident. As the day went on, the crisis team provided the investigators with any additional information they requested.

At the end of the day, Dr. Jones addressed the crisis team, "Thank you for your cooperation. We'll be here for the next month conducting our investigation. We'll keep you informed of our progress and any new developments."

Mr. Macguire thanked the investigators for their diligence and professionalism, "We appreciate your efforts and look forward to working with you throughout this investigation." The crisis team nodded in agreement, and the investigators departed for the day.

The crisis team reconvene for a meeting

After the TGA investigators left, the Quintomo Hospital crisis team reconvened to discuss the developments and plan the way forward. The team included the CEO, Patrick Macguire, the Chief AI Officer, Bill Stone, the Chief Technology Officer, Jane Lee, the Director of Cybersecurity, Julia Stephanopoulos, and the Director of Security, Alan Palmer.

Patrick Macguire opened the meeting, "We've had an intense few weeks with the TGA investigation, staff shortages, and the hiring process. The good news is we have completed hiring, and the new staff will be in the hospital within 4 weeks."

Bill Stone added, "So far our internal review reveals that we have followed best practices in our security protocols. We continue to review all the data and run simulations of other scenarios to stress test our systems."

Julia Stephanopoulos interjected, "Indeed, we need to ensure we continue to follow best cybersecurity practices to avoid future incidents."

Jane Lee added, "I'm providing the required access to CyberPower, the cybersecurity firm doing an independent review of our cybersecurity."

Vladimir Kushnir spoke up, "We also need to continue with our training and upskilling programs for staff to ensure they are comfortable with the new technologies and can handle situations when they arise."

Patrick Macguire added, "We also need to focus on rebuilding morale within the hospital staff. The recent events have taken a toll on them, and we need to show our appreciation and support."

The team then discussed their plans for an Employee Appreciation Day and other initiatives to boost morale. They also discussed increasing communication and transparency with the staff to ensure they were kept up to date on the hospital's progress.

A new lead about the perpetrators

As the crisis meeting at Quintomo Hospital continued, Primeo announced that it was time for the scheduled call with Clyde Boston, the government's security agent. The team braced themselves for the latest update on the ongoing investigation into the cyberattack and the robot malfunction that led to the tourists' injuries.

Clyde's voice came through the speakers, "Good evening, Mr Macguire and the rest of the crisis team. As I mentioned earlier, we are cooperating with Interpol and other foreign security agencies in tracking the suspects behind the cyberattack and the robot malfunction. We have received a lead from the CIA and we look forward to a meeting with their operatives tomorrow to get more details."

Mr Macguire nodded, "Thank you for the update, Clyde. We appreciate your hard work on this case, and we look forward to hearing more tomorrow."

As the call ended, the crisis team let out a collective sigh, the tension in the room lifting slightly with the news of progress in the investigation. But they knew that there was still a long way to go, and they needed to stay focused and vigilant.

As the meeting came to an end, Mr Macguire addressed the team, "Thank you all for your hard work and dedication to Quintomo Hospital. I know this has been a trying time for all of us, but we must continue to work together to ensure the safety and well-being of our patients and staff."

The team nodded in agreement, and they left the meeting room with a renewed sense of purpose, ready to face whatever challenges lay ahead.

Chapter 15

TRACKING THE PERPETRATORS

The cyberattack on Quintomo Hospital sent shockwaves throughout Australia and the world, propelling a full-scale investigation into action. Clyde Boston, the Australian lead investigator, was hell-bent on locating the perpetrators and delivering justice. He spearheaded a team of experts, each bringing unique skills and expertise to the table.

Samantha, the cybersecurity expert, was assigned the task of analysing the data and identifying the vulnerabilities in the hospital's system. Jack, the data analyst, was to sift through volumes of information for patterns or anomalies. Sarah, the forensic accountant, tracked the financial trail of the cyber criminals, while Emma, the legal counsel, ensured the evidence gathered would stand up in court. Tim served as the liaison officer, coordinating efforts and communication with other agencies.

Joining the Australian team were two operatives from the Central Intelligence Agency (CIA) of the United States: Mike Sunderland and Audrey Salt. The pair brought years of intelligence on the cyber criminals to the investigation. Mike helmed cyber operations, while Audrey was his deputy. This team of experts vowed to leave no stone unturned in their quest to bring the cybercriminals to justice.

As the virtual meeting commenced, the faces of the two CIA operatives appeared on the screen. The Australian team sat on the edge of their seats, eager for the CIA's insights. Mike Sunderland began with an introduction of himself and Audrey Salt before delving into the CIA's investigation.

"Our investigation into this trio began a few years ago, following suspicions of their involvement in cyberattacks on US-based corporations," Sunderland explained. "We've been tracking their movements and activities ever since."

Salt interjected, "We've amassed considerable information on these three individuals, including their methods and techniques. They typically operate as fee-for-service operatives, working for anyone willing to pay their high prices. Their associations span nation-states, criminal organizations, and even corrupt corporations."

"Do you have any information on their recent activities leading up to the attack on Quintomo Hospital?" Samantha, the cybersecurity expert, asked.

"No, we have no leads from the past six months. They've been under the radar. Prior to this, their base of operation was in Eastern Europe where they coordinated a high-profile cyberattack on Eurofuture Bank," Sunderland responded.

"Do you have any ideas about their current locations?" Jack, the data analyst, asked, pulling up a map on the screen.

"We believe they're currently in either Southeast Asia or Eastern Europe. We're collaborating with Interpol and foreign security agencies to track their movements," Sunderland answered.

Emma, the legal counsel, chimed in, "Do you have any information on their clients or associates?"

Salt nodded, "We've compiled a list of potential clients and associates. We suspect they're well-connected and have access to powerful individuals and organizations. We'll share this information with you."

As the meeting continued, tension in the room escalated. The Australian team was grateful for the CIA's cooperation, but the urgency of the situation weighed heavily. The longer the trio remained at large, the greater the risk to national security.

The meeting ended after an hour-long presentation, leaving the Australian team with a wealth of information to digest and act upon.

Debrief and follow-up

"We need to act fast," Clyde insisted. "Samantha, Jack, and Sarah, analyse the IP addresses and potential locations we've been given. Emma, handle the legal aspects of the investigation. Tim, liaise with Interpol and foreign security agencies."

The team members exchanged glances, a whirlwind of thoughts whirring in their minds. Clyde broke the silence, summarizing key takeaways from the meeting. Each possibility was considered, from a straightforward ransomware attack to the theft of intellectual property, to the involvement of a nation-state actor. Clyde even suggested the possibility of an insider threat, though without specific suspects, this theory was purely speculative.

"We can't rule anything out," Clyde asserted. "It could even be a message or a warning to Quintomo or another organization."

They all agreed to maintain open-mindedness about the motives behind the attack. "The more angles we consider, the better chance we have of identifying the perpetrators and bringing them to justice," Clyde added.

Turning back to the information gleaned from the CIA, Clyde identified their next steps. "We need to analyse the IP addresses and potential locations. We also must investigate the trio's past attacks and their possible associates. We should scrutinize any financial transactions linked to them."

Each team member had clear directives. Samantha and Jack would analyse technical data. Sarah would probe financial links, with Emma handling legal requirements. Clyde emphasized the need for urgency, scheduling an update meeting within a few days.

Discussing ongoing collaboration with the CIA, the team decided on regular virtual meetings to exchange updates. Clyde would serve as the primary contact for the Australian team, with Mike Sunderland fulfilling that role for the CIA. The team even considered the possibility of sending members to the U.S. for short-term assignments to facilitate collaboration.

Finally, the team considered potential challenges and risks. Emma pointed out the risk of alerting the trio, who might then go into hiding or retaliate. Clyde acknowledged this, underscoring the need for maintaining their cybersecurity and vigilance. The team agreed to prioritize tracking the trio and securing their own systems.

With the meeting concluded, the team was filled with determination. Clyde's final words rang in their ears, "We have a lot of work ahead, but I have faith in this team. Let's stay focused and keep pushing forward."

The implications of the investigation were significant, reminding them of the growing risks posed by cyber threats in the advancing digital world. It highlighted the need for strong cybersecurity measures, information sharing, and international collaboration to prevent such attacks.

"It is imperative that we continue to learn from such incidents and take steps to prevent and mitigate the impact of cyber threats," Clyde concluded.

Meanwhile, more crisis loomed at the hospital as staff continued to work under immense pressure without the collaboration of the humanoid robots.

Chapter 16

Nurses in Crisis: Overcoming Challenges at Quintomo Hospital

As Joan Johnson, a retired nurse who recently returned to help at Quintomo Hospital, made her way through the Rehabilitation ward, she couldn't help but feel overwhelmed by the chaos. The suspension of the humanoid robot program had caused a lot of disruptions to the daily routines of the nurses, patients, and other staff members.

She walked up to Tony Burke, a young nurse who had cut his teeth working at Quintomo hospital with robot assistance. "Tony, how are you holding up?" she asked.

"It's been tough, Joan," Tony replied. "We're having to be more hands-on with patients as we no longer have robots to help with patient transfers and other patient care services such as transportation and special watch for agitated or delirious patients.

It's a lot of extra work, and we've been stretched thin trying to keep up."

Joan nodded in agreement. She knew all too well how difficult it could be to provide quality care to patients without the aid of technology. However, she was also heartened by the new recruits from African countries, Southeast Asia, and Pacific countries who were now working alongside her and Tony.

They were excelling in practical matters that used their old-fashioned nursing skills while struggling to rely on technology, including Primeo AI assistant. Despite a rocky start, they quickly found their footing and were adding value to the hospital.

However, things were still slower without the robotic assistance, and the hospital was struggling to keep up with the demand for nursing care.

A day in the life of nurse Joan Johnson

Joan Johnson was a 65-year-old retired nurse who had worked in the intensive care unit at the Royal Melbourne Hospital before retiring. She had recently returned to work to help with the staff shortage caused by the suspension of the humanoid robot program. Joan found the changes in the hospital to be overwhelming, especially since she was not very tech-savvy. However, she was determined to help in any way she could.

One day, Joan was working on the rehabilitation ward when she noticed that one of her patients, an elderly man named George, was becoming increasingly agitated. Joan tried to calm him down, but he was not responding to her usual techniques. Normally, the robots would have been used to monitor and calm agitated patients, but with their suspension, Joan had to come up with a new solution.

Joan tried to calm George down with a soft voice and reassuring words, but he continued to groan and flail his arms. She then referred to his beautiful family photo on the wall and adjusted his blankets to make him more comfortable. When that didn't work, she tried a different approach and asked if he wanted to watch TV, but he just shook his head and groaned louder. Frustrated, Joan was about to give up when she heard Primeo's voice in her earpiece. "Joan, I've pulled up George's profile and it says he loves classical music. Why don't you try playing some for him?" Joan quickly accessed the music library and selected a classical playlist. As soon as the music started, George's movements slowed down and his breathing became more even. He closed his eyes and a peaceful expression spread across his face. Joan couldn't believe the difference the music had made and thanked Primeo for the suggestion. Joan continued to play the music for him throughout the day, and he remained calm and relaxed.

As Joan was leaving the hospital that day, she saw George's family waiting in the lobby. They thanked her for her help with their father and were amazed at how she had managed to calm him down. They had never seen him so relaxed.

Joan felt a great sense of pride in her work that day. She realized that even without the help of the robots, she was still able to make a difference in the lives of her patients. She was grateful to Primeo for coming to her aid. She knew that there were still many challenges ahead, but she felt a renewed sense of purpose and determination to help in any way she could.

A day in the life of nurse Tony Burke

Tony Burke felt the weight of his responsibilities as he entered the room of his patient, Zack, who had recently suffered a stroke. Normally, with the help of the humanoid robots, transferring Zack would be a breeze. But now, Tony had to rely on his own strength and skill to safely move Zack from the bed to a wheelchair.

Tony positioned himself beside the bed, ready to help Zack up. "Okay Zack, let's take it slow and easy," he said, gently guiding the patient's arms around his own shoulders. Together, they slowly stood up, but Tony felt the strain in his back as he took most of Zack's weight.

As they began the transfer, Tony suddenly felt Zack's weight shift, and the patient began to fall. With a quick reaction, Tony adjusted Zack's position, so he landed on his buttocks instead of hitting his head. He let out a sigh of relief as he helped Zack back up.

"I'm sorry about that, Zack," Tony said, feeling flustered. "I'll be more careful next time."

Despite his best efforts, Tony continued to feel apprehensive throughout the day. He carefully assisted Zack with his other transfers, but he couldn't shake the feeling of uncertainty.

At the end of his shift, Tony sat down in the break room, feeling sad and defeated. "I don't know how I'm supposed to do this without the robots," he muttered to himself.

But just then, Primeo spoke up. "Tony, you did an excellent job today. I know it's a tough adjustment without the robots, but you're doing great."

Tony looked up, surprised at the praise from the AI assistant. "Thanks, Primeo. I appreciate that."

"Perhaps tomorrow we can review some techniques to make the transfers easier and safer for both you and the patients," Primeo suggested.

Tony felt a glimmer of hope. Maybe with some more practice and guidance, he could become more confident in his manual transferring skills. He left the hospital that day feeling determined to rise to the challenge and be the best nurse he could be, robots or no robots.

A day in the life of nurse Ngozi

Ngozi, a newly recruited nurse from Nigeria, had recently joined the dedicated team at Quintomo Hospital in Australia. Though she faced some initial challenges adjusting to the new culture and healthcare system, she had quickly gained recognition for her compassionate care and diligent work ethic. Patients valued her presence, and her colleagues held her in high regard.

On a particular day, Ngozi was assigned to care for Mr. Smith, a post-surgery patient who had undergone a hip replacement. Mr. Smith was enduring significant pain and discomfort, and Ngozi was determined to assist him in managing his symptoms effectively.

As Ngozi prepared to administer Mr. Smith's pain medication, she discovered an issue with the dosage. The medication required delivery through a sophisticated high-tech machine designed to regulate the dosage automatically, but it seemed to be malfunctioning.

Recognizing the urgency of Mr. Smith's situation, Ngozi sought assistance from her experienced colleague, Angie. With a friendly smile, Angie empathized with Ngozi's predicament and swiftly devised a solution.

"Hey, Ngozi! Don't worry. I know where we can get a new machine," Angie said reassuringly. "There's a supply room on the third floor where they keep extra equipment. Let's go and find a replacement for Mr. Smith."

Grateful for Angie's guidance, Ngozi eagerly followed her colleague to the supply room. Angie helped her navigate the hospital's layout, pointing out the correct corridor and explaining where to find the equipment they needed.

Once they arrived, Angie swiftly located a brand-new machine that matched the specifications required for Mr. Smith's pain medication. She guided Ngozi through the process of setting it up, ensuring that everything was connected properly and functioning smoothly.

Within minutes, Ngozi had the new equipment up and running. She marvelled at the efficiency of her colleague and thanked Angie profusely for her invaluable assistance.

With the new machine in place, Ngozi could now administer the pain medication accurately and precisely for Mr. Smith's needs. Together, Ngozi and Angie monitored Mr. Smith closely, observing his vital signs and providing the necessary care and support.

Mr. Smith received the correct dose of pain medication at the right time, thanks to the collaborative effort between Ngozi and Angie. Their quick thinking and problem-solving abilities had a positive impact on the patient's recovery.

Ngozi felt a deep sense of gratitude for Angie's kindness and expertise. Their teamwork not only ensured excellent patient care but also served as a reminder of the supportive environment within the hospital. The experience reinforced Ngozi's confidence and reaffirmed her belief in the strength of teamwork and camaraderie among healthcare professionals at Quintomo Hospital.

At the end of their shift, Ngozi and Angie walked out of the ward together. Ngozi turned to Angie and said, "Thanks again for your help earlier with the medication administration device. I couldn't have done it without you."

Angie smiled warmly at Ngozi and replied, "No problem at all. You did a great job handling it. Good job!"

Ngozi felt a sense of relief and gratitude, knowing that she had made a positive impression on her colleague. As they continued walking, Angie turned to Ngozi and said, "By the way, I have to say, your English is really good."

Ngozi smiled back and replied, "Thank you. English is the official language of Nigeria, and I grew up speaking it. I even went to school in English."

Angie looked surprised and impressed. "Wow, I didn't know that" she said. "Well, welcome to Australia. We're glad to have you here."

Ngozi felt a sense of belonging and acceptance in her new home. She knew there would be challenges ahead, but with the support of her colleagues like Angie, she felt confident that she could overcome them.

Sarah and Kate, registered nurses, assist a team member with bed mobility

As Sarah and Kate made their way through the hospital, they caught sight of a scene in the corridor that caught their attention. A new registered nurse from the Philippines, Maria, and a patient care assistant named Scott were struggling to wheel a patient's bed from the theatre back to the ward. The bed seemed to be jammed against the door, leaving Maria perplexed.

Noticing their predicament, Sarah approached them with concern, asking, "Do you need any assistance?"

Maria sighed in relief, grateful for the offer. "Yes, please. I can't figure out what's wrong. The bed won't budge."

Kate, always ready to problem-solve, stepped in with a suggestion. "Let's try pushing it together on the count of three, Maria and Scott. Perhaps we can dislodge it."

In unison, they synchronized their efforts, pushing with all their might, but the bed stubbornly refused to move. Frustrated, they examined the bed more closely, hoping to find a solution. It was then that Maria realized she had inadvertently turned off the motorized bed without switching to manual mode.

Sarah, ever observant, quickly assessed the situation. "Wait a moment. I think I've got it. Let's try restarting it."

Following Sarah's lead, they restarted the bed. With a flicker of hope, they turned on the automatic self-driving mode. To their delight, the bed sprang to life, smoothly gliding forward.

Relieved, Maria expressed her heartfelt gratitude to Sarah and Kate for their assistance. Together, they carefully guided the bed into the lift and continued transporting the patient back to the ward effortlessly.

As they walked out of the hospital, a sense of camaraderie filled the air. Kate reflected, "It's always rewarding to lend a helping hand, especially when unexpected challenges arise. We're here to support one another."

Sarah nodded in agreement, a smile gracing her face. "Absolutely. It's reassuring to know that we can rely on our colleagues as a team. We're stronger together."

With a renewed sense of unity and shared purpose, they continued their journey, ready to face any future obstacles knowing that their support and teamwork would always make a positive difference in providing exceptional patient care.

A day in the life of nurse Simon

Simon had been working at Quintomo Hospital for just over a year, having graduated from nursing school with top honours. He had always been passionate about healthcare and was eager to learn as much as he could on the job. Today was no different, except that he had been assigned to care for a patient who was proving to be quite challenging.

The patient, Jim, a retired 72-year-old boxer, had been admitted to the General Medicine ward with a urinary tract infection that had become complicated by delirium. He required constant special watch and was often agitated, making it a challenging situation for Simon to balance his care and judiciously use sedatives when needed.

It was total chaos as Jim landed a powerful punch on Simon's face during an episode of severe agitation. Simon stumbled back and collapsed to the ground, unconscious. The other nurses rushed to his aid, trying to revive him while also trying to contain the patient. Thankfully, he had a pulse and his vital signs turned out to be normal. The ward was in disarray as an emergency code was called, and within minutes, four security officers arrived to help contain the situation.

The security officers took control of the patient, another code had been also called for Simon's loss of consciousness. The emergency team rushed in, taking Simon to the emergency department for urgent medical attention. His fellow nurses looked on in shock, hoping that he would pull through.

Meanwhile, the nursing director, Jen Nguyen, was notified of the situation and arrived on the scene to assess the situation. The patient was placed under additional monitoring, while Simon underwent a series of tests in the emergency department. The situation was tense, and nobody knew what the outcome would be.

Nim, a robocaller, rings Simon's girlfriend

As Sonia sat on her couch watching TV, her phone began to ring. She hesitated for a moment, not recognizing the number, but then decided to answer.

"Hello?" she said, slightly apprehensive.

"Good evening, this is Nim, a robocaller from Quintomo Hospital. Is this Sonia?" the robovoice responded.

Sonia's heart rate increased as she listened to the robot's introduction. She had a feeling it was related to Simon, so she replied, "Yes, this is Sonia. What's happened? Is everything okay?"

Nim continued, "I am sorry to inform you that Simon had an accident at the hospital today. He was punched by a patient and lost consciousness. He was immediately taken to the emergency department for treatment."

Sonia's mind was racing with worry and fear as she processed the information. "Oh my God! Is he okay? What's his condition?" she asked, desperation lacing her tone.

"Nurses and doctors are currently assessing him in the emergency department, but he is still unconscious but has stable vital signs. He will soon be getting a CT scan of his brain done." Nim assured her.

"Can I come see him?" Sonia asked, already standing up and heading to the car.

"Yes, you can visit him. He is currently in the resuscitation bay 5," Nim replied.

"Okay, thank you. I'm on my way," Sonia said before ending the call and rushing out the door.

Chapter 17

THE BATTLE CONTINUES: HOPE AND SUPPORT FOR MATT BARRY'S RECOVERY

As the dust settled on the cyberattack incident at Quintomo Hospital, the hospital staff and executives took a deep breath and looked back on the past month. The crisis team had worked tirelessly to ensure that the hospital continued to function at the highest level, despite the challenges posed by the attack. The hospital had also undertaken a massive public relations (PR) campaign to rebuild public trust and confidence.

At 7 am, the crisis team assembled in the situation room. Patrick Macguire, the CEO of Quintomo Hospital, Prof. Bill Stone, the Chief AI Officer, Dr. Jane Lee, the Chief Technology Officer, Julia Stephanopoulos, the Director of Cybersecurity, Alan Palmer, the Director of Security, and Vladimir Kushnir, the Deputy Director of Cybersecurity were all present. The meeting started with the CEO commending the team's hard work in dealing with the aftermath of the cyberattack incident.

Stephanopoulos gave an update, indicating that the hospital had not suffered any further attacks and that there were no indications that an attack was imminent. She mentioned that they were still waiting for the report from CyberPower and the Therapeutic Goods Administration (TGA) TGA, which were due in a couple of weeks.

Dr. Jane Lee mentioned that the cyber defence systems of the hospital had been updated with the learnings from the incident. She also highlighted the need for ongoing monitoring and updates to ensure that the hospital remained secure.

Prof. Bill Stone then spoke up, addressing the impact of the suspension of the humanoid robot program to the hospital at large. He mentioned that the staff had stepped up to fill the gaps left by the robots, but there were still some challenges to overcome. He suggested that they could explore alternative solutions to supplement the staff and ease the burden.

Alan Palmer, the Director of Security, raised concerns about the security protocols of the hospital's third-party vendors. He suggested that they should conduct a review of their security measures to ensure that they were up to par with the hospital's own standards.

Vladimir Kushnir, the Deputy Director of Cybersecurity, added that they should also continue running regular security awareness programs for the hospital's staff.

The meeting continued with the team reviewing updates and discussing potential solutions. The mood was lighter than before, with a sense of relief that they were slowly getting back to normalcy.

Quintomo Hospital's PR campaign

Fiona White, the Chief Operating Officer, joined the meeting virtually and shared her screen to present the latest developments in the hospital's PR campaign. "Thank you all for your hard work and dedication over the past month," she began. "We have made significant progress in our efforts to improve public sentiment towards Quintomo Hospital."

Fiona then outlined the various PR interventions they had undertaken. "We started with television adverts featuring testimonials from patients and their families about the excellent care they received at Quintomo Hospital," she said. "We also arranged for our scientists and researchers to appear on TV shows and podcasts to discuss their groundbreaking work. Our cooperation with a documentary about Quintomo Hospital turned out well, and it received positive reviews."

Fiona then added two additional PR strategies that were employed. "We have also increased our presence on social media platforms to engage with our audience and address any concerns they may have," she said. "Lastly, we organized a series of events for the public, including free health screenings and seminars on various health topics."

Patrick Macguire, the CEO of Quintomo Hospital, leaned forward in his seat. "Please tell us more. How have our efforts been received?"

Fiona cleared her throat and began to read from her slides. "We've been monitoring social media sentiment across multiple platforms, and I'm happy to report that the overall sentiment towards Quintomo has improved by 25%. We're also seeing positive results in our patient satisfaction surveys, which have gone up by 10% compared to last month."

The room erupted in applause, and Fiona smiled.

"That's great news," said Prof. Bill Stone, the Chief AI Officer. "It's a testament to the hard work of everyone involved in the PR campaign."

Fiona nodded in agreement. "It's been a team effort, and I'm grateful to everyone who has contributed."

Stephanopoulos, the Director of Cybersecurity, cleared her throat. "While it's encouraging to see the sentiment towards the hospital improve, we can't let our guard down. We need to remain vigilant and continue to strengthen our cyber defences to ensure that we don't experience another attack."

Alan Palmer, the Director of Security, nodded in agreement. "Absolutely. We've also been working on increasing our physical security measures in case of any further incidents."

Vladimir Kushnir, the Deputy Director of cybersecurity, added, "We're also waiting for the reports from CyberPower and TGA. Once we have those, we can further enhance our security measures and ensure that we're not vulnerable to any future attacks."

Fiona listened intently to the conversation before interjecting, "I agree. But I also want to stress the importance of maintaining a positive image and reputation for the hospital. It's not just about cyber defence, but also about the trust and confidence that our patients and the public have in us."

The team nodded in agreement, and the meeting continued with a discussion on how to further improve public sentiment towards Quintomo Hospital.

Director of nursing updates executives on nursing crisis

Just as the discussion was concluding, Primeo announced to Mr. Macguire that he had received a call from Jen Nguyen, the Director of Nursing. Jen was patched through and told the team about the horror story of how a delirious patient had punched Simon, a registered nurse on the medical ward, in the face, knocking him unconscious. She went on to express her frustration with the ongoing suspension of the humanoid robot program and how Simon's injury would not have happened if the robots were in operation.

Patrick Macguire asked pertinent questions about the incident, then indicated that he would go right away to the Emergency Department to see Simon and asked Bill Stone and Jane Lee to accompany him, to which they agreed.

As they walked towards the elevator, Mr. Macguire turned to Dr. Lee and said, "We need to do something about this suspension of the humanoid robot program. It's putting our staff and patients in danger."

"I agree," replied Dr. Lee. "We need to convince the board of directors to reconsider the decision. The benefits that the robots bring far outweigh any risks."

Prof. Bill Stone chimed in, "I can provide data on how the robots improve patient outcomes and reduce the workload on our staff. I think that would be persuasive to the board."

As they arrived at the Emergency Department, they saw Simon sitting up in bed, looking groggy. Mr. Macguire went over to him and asked how he was feeling.

"I'm okay, just a little sore," replied Simon. "I don't remember much after the patient punched me."

Dr. Lee spoke up, "We're going to make sure you get the best care possible, Simon. We're also going to work hard to get those robots back online as soon as possible so that incidents like this don't happen again."

Simon smiled weakly, "Thanks, that would be great." His girlfriend, Sonia, who was there with him also thanked them.

The team left Simon to rest and walked out of the Emergency Department, deep in conversation about how to convince the board of directors to reconsider the decision to suspend the humanoid robot program. They knew it would be an uphill battle, but they were determined to do everything they could to improve patient care and safety at Quintomo Hospital.

Matt Barry's high-profile case

As the CEO and his team left the emergency department, the sounds of alarms started blaring, indicating a code stroke. The emergency department was in chaos as Primeo announced an estimated time to arrival of 5 minutes. The high-profile patient, Matt Barry, a member of parliament from New South Wales, was on holiday in Quintomo. He was walking along the Flinders trail with his wife when he noted the worst headache of his life, vomiting, drooping of his right eyelid, and difficulty with word-finding. The medical team raced against time to save his life and prevent any long-term damage.

Matt Barry, the high-profile politician, was rushed into the hospital on a stretcher by the paramedics, his face contorted in pain. The emergency department was in a flurry of activity as the team of doctors and nurses sprang into action to assess his condition. After initial assessment, he was taken directly to the radiology department to undergo an MRI and MR angiogram.

Dr. Jess Lewis, the emergency physician on duty, reviewed the imaging results and her face tensed up as she saw the images of Matt's brain. She took a deep breath and walked into the resuscitation bay where Matt Barry lay with his wife seated by his side.

"Mr. Barry, we have reviewed your imaging results, and I'm afraid I have some difficult news to share," Dr. Lewis said, her voice calm and reassuring.

Matt sat up in his hospital bed, his expression anxious. "What is it, doctor?" he asked.

"The MRI and MR angiogram show that you have a posterior communicating artery aneurysm, which is a bulging weak spot in a blood vessel in your brain," Dr. Lewis explained. "It's a serious condition that has led to your symptoms and our algorithm predicts a high risk of bleeding within 24 hours."

Matt's face paled as he took in the gravity of the situation. "What are my options?" he asked, his voice barely above a whisper.

Dr. Lewis proceeded to explain the available treatment options to him, including the risks and benefits of each. Matt listened intently, his mind racing as he tried to process the information.

At the end of the conversation, Dr. Lewis reassured him that they would do everything in their power to ensure his safety and comfort during his stay at Quintomo Hospital.

Media report on Matt Barry's hospitalization

At the news studio, Global News Media anchors, Tim and Jo, were reporting on the breaking news of Matt Barry's hospitalization. "Good evening, we are interrupting our regular programming to bring you breaking news," Tim began. "Member of Parliament from Urbano, New South Wales, Matt Barry, has been admitted to Quintomo Hospital after suffering a medical emergency," Jo added.

As they spoke, reporters from various media outlets gathered outside the hospital, including Lisa from ABC News, Tom from Channel Nine, and Sophie from The Australian. Lisa started her live broadcast, "We're here outside Quintomo Hospital where Matt Barry, a high-profile politician, has been admitted after suffering a medical emergency."

Tom jumped in, "We don't have all the details yet, but sources tell us that he may have suffered a stroke." Sophie interjected, "That's not confirmed, Tom. We have heard rumours that it could be something else entirely."

Back in the studio, Tim and Jo were discussing the situation. "It's important to note that we don't have all the facts yet," Tim said. "Absolutely, Tim. We are still waiting for an official statement from Quintomo Hospital," Jo added.

Meanwhile, outside the hospital, the reporters continued to speculate. "I've heard that it could be a heart attack," Lisa said. "Or maybe it's a brain tumour," Tom added. Sophie shook her head, "We shouldn't jump to conclusions. We need to wait for official information."

Back in the studio, Tim and Jo signed off, "We will continue to bring you updates on this breaking news as we receive them. Our thoughts are with Matt Barry and his family at this time."

Well wishes to Matt Barry

As news of Matt Barry's condition spread, messages of support poured in from politicians, celebrities and the public alike. His political colleagues rallied around him, sending well wishes and hoping for a speedy recovery.

The Prime Minister, Amy Roberts, released a statement saying, "Matt is a valued member of our team, and we are all thinking of him and his family during this difficult time. We wish him a full and speedy recovery."

The Opposition Leader, Sarah Jones, echoed the sentiment, saying "Our thoughts are with Matt and his family. We are sending our well wishes for a quick recovery."

Celebrities also joined in to send their support, with actor Hugh Cummings tweeting, "Sending love and positive vibes to Matt Barry and his family. Get well soon!"

Social media was also flooded with messages of support, with the hashtag #GetWellMatt trending across the country.

The outpouring of support was overwhelming, and it showed the unity of the Australian people in times of crisis.

Matt Barry consents to an interventional procedure

John and Bev rushed to Quintomo Hospital after they received news of their father's hospitalization. They found their mother, Vicki, standing outside Matt's room. She looked worried and pale.

"How's dad doing?" John asked, putting a comforting hand on his mother's shoulder.

Vicki shook her head. "They found a brain aneurysm that has a high risk of rupture and bleeding within the next 24 hours."

Just then, a nurse walked out of Matt's room. "Are you family?" she asked.

"Yes, we are," Bev replied.

The nurse nodded. "The doctor will be in shortly to speak with you. Your father is scheduled for a catheter-based procedure later this evening."

John frowned. "What kind of procedure?"

"It's called an interventional neurology catheter-based coiling procedure," the nurse explained. "It's meant to block blood flow from the aneurysm and prevent it from rupturing."

John and Bev exchanged worried glances. They knew their father was in good hands, but the news still made them anxious.

Inside the room, Matt was lying in bed, looking pale and tired. When he saw his children, a weak smile formed on his lips. "John, Bev, it's good to see you."

"Dad, what happened?" Bev asked, tears forming in her eyes.

Matt took a deep breath. "I had a headache and some other symptoms, so I came here to get checked out. They found an aneurysm."

John frowned. "What's that?"

"It's a bulge in one of the arteries in my brain," Matt explained. "It can be dangerous if it ruptures, but they're going to fix it tonight."

Vicki nodded. "The doctors here are the best. You're in good hands, honey."

Matt smiled weakly. "I know. I'm just glad you're all here with me."

Later that evening, the doctor arrived to discuss the procedure with Matt and his family. He explained the details of the coiling procedure and answered their questions. Dr. Paul Chen, Interventional Neuroradiologist, sat down with Matt Barry to explain the endovascular coiling procedure. "Mr. Barry, we will insert a catheter into your femoral artery and guide it to the site of the aneurysm in your brain. We will then use tiny platinum coils to pack the aneurysm and prevent it from rupturing."

Matt Barry looked concerned and asked, "What are the risks associated with this procedure?"

Dr. Chen replied, "Firstly, one of the most common complications is a thromboembolism. This occurs when a blood clot forms and can block blood flow to the brain, leading to a stroke. Another possible complication is a rupture of the aneurysm during the procedure, which can cause bleeding in the brain.

In addition, there is a risk of vessel perforation, which means a small hole could be made in the blood vessel during the procedure. This can cause bleeding and increase the risk of a stroke. Another possible complication is vasospasm, which is a narrowing of the blood vessels and can cause reduced blood flow to the brain, leading to a stroke.

It is important to note that while these complications can occur, they are rare and the benefits of the procedure often outweigh the risks. Our team will take every precaution necessary to minimize the risk of complications and ensure your safety during the procedure. If you have any questions or concerns, please do not hesitate to ask."

Matt Barry took a deep breath and said, "Okay, I understand. I'm ready to proceed with the procedure."

Dr. Chen smiled reassuringly and said, "Great, thank you for your trust. We will take good care of you."

Matt Barry gave his voice consent by saying, "I accept" and was taken into the interventional neuroradiology suite for the procedure.

Dr. Chen smiled reassuringly and said, "Great, thank you for your trust. We will take good care of you."

Chapter 18

ROAD TO RECOVERY

Matt Barry was in the surgical ward, surrounded by his family. Dr. Paul Chen, the interventional neuroradiologist, came in to speak with them. He pulled up the pre and post images from the procedure on his tablet and began to explain.

"I'm happy to report that the aneurysm has been successfully coiled," Dr. Chen said. "We were able to block off the blood flow to the aneurysm, which should prevent it from rupturing in the future."

Matt smiled, relieved. "That's great news," he said. "I'm feeling much better already."

Dr. Chen continued, "We were also able to observe that the aneurysm was quite large, and the risk of bleeding was high. But with the coiling procedure, the risk has been significantly reduced."

Matt's daughter, Bev, breathed a sigh of relief. "Thank goodness," she said. "We were all so worried."

Dr. Chen nodded. "We'll continue to monitor Mr. Barry closely, but I'm optimistic that he will make a full recovery. The next step is for him to rest and recover here in the hospital for a few more days."

Matt's wife, Vicki, reached out to hold his hand. "Thank you so much for everything you've done for us," she said to Dr. Chen.

"It's my pleasure," Dr. Chen replied. "I'm just glad we were able to help."

As Dr. Chen turned to leave the room, Matt Barry mentioned the recent incident at the hospital and enquired how he and other hospital staff were holding up.

Dr. Chen nodded and replied, "Yes, it has been a trying time for all of us here at Quintomo Hospital. But we have a strong team and we have been working hard to overcome the challenges."

Matt Barry looked up at Dr. Chen and said, "I want to thank you and your team for your hard work and dedication. Without you, I wouldn't be here today."

Dr. Chen smiled and replied, "It was our pleasure to help you, Mr. Barry. We are just doing our job."

Matt's son John spoke up, "Dad, we've been keeping up with the news about the hospital. It's been all over the media."

Matt nodded and replied, "Yes, I know. It's a shame that the good work of this hospital is being overshadowed by the cyberattack."

His daughter Bev added, "But we've also seen a lot of support and well wishes from the public. People know that the hospital staff are doing their best under difficult circumstances."

Matt smiled and looked at his family. "I'm lucky to have such a supportive family by my side. And I'm grateful to the hospital staff for their excellent care."

The family continued to chat for a while, discussing the recent events at the hospital and the plans for Matt's recovery. The mood was positive and optimistic, with everyone feeling grateful for the successful outcome of the procedure.

Matt Barry is discharged home

Matt Barry stood in front of Quintomo Hospital, flanked by his wife Vicki and his children John and Bev. Behind them, the CEO of Quintomo Hospital, Patrick Macguire, and Dr. Paul Chen, the interventional neuroradiologist who performed the coiling procedure on Matt, stood at a respectful distance.

As cameras flashed and reporters jostled for position, Matt cleared his throat and began to speak. "I just want to start by saying how grateful I am to my family for their unwavering support during this difficult time. I also want to thank the incredible doctors and nurses at Quintomo Hospital for their expertise and care. I truly believe that their quick actions saved my life."

Matt paused for a moment, looking out at the crowd of reporters and well-wishers. "Finally, I want to thank the public for their overwhelming support and concern. It's been truly heartwarming to see so many people come together during a difficult time."

As Matt finished his statement, the reporters began shouting questions at him. Megan, the reporter from Global News Media, pushed her way to the front of the crowd. "Mr. Barry, can you tell us more about your recovery? How are you feeling now?"

Matt smiled. "I feel great, thank you. My headache is gone, and my speech and vision have returned to normal. I couldn't have asked for a better outcome."

Megan nodded, taking notes. "And what about the recent cyberattack on the hospital? Do you have any thoughts on that?"

Matt's expression turned serious. "It's a sobering reminder of how important it is to have strong cybersecurity measures in place. I'm grateful to the hospital for their quick response and for keeping me safe during my stay."

Patrick Macguire stepped forward, offering his own thoughts on the matter. "We're taking steps to strengthen our cybersecurity protocols, and we're confident that we can prevent any future attacks. The safety and wellbeing of our patients is always our top priority."

Dr. Paul Chen added his own comments, describing the procedure he performed on Matt and the importance of early intervention in cases like his. As the reporters continued to shout questions and take pictures, Matt and his family prepared to head home, grateful for the care they received at Quintomo Hospital.

The recovery of Matthew Clark

Matthew Clark had a long road to recovery having been on the rehabilitation program for over a month. After waking up from his coma in the intensive care unit (ICU), he was referred to a rehabilitation program to help him regain his strength and improve his cognitive function. The rehabilitation program consisted of physical therapy, occupational therapy, and speech therapy.

Matthew's physical therapy focused on regaining his strength and mobility. His occupational therapy helped him learn and practice activities of daily living, such as getting dressed and feeding himself. His speech therapy focused on helping him regain his ability to communicate and to swallow safely.

At first, Matthew struggled with the rehabilitation program. He was weak and disoriented, and he had trouble remembering things. But with time and dedication, he slowly made progress. He began to regain some of his strength and coordination, and his memory started to improve.

The rehabilitation team worked closely with Matthew to develop a personalized treatment plan that addressed his specific needs and goals. They encouraged him to set small, achievable goals for himself and to celebrate his progress along the way.

Matthew's family played a crucial role in his recovery. They visited him regularly and provided him with emotional support and encouragement. They also worked closely with the rehabilitation team to ensure that Matthew was receiving the best possible care.

Over time, Matthew made significant progress in his rehabilitation. He regained much of his strength and coordination, and his memory and cognitive function improved significantly. With the support of his family and the rehabilitation team, he was able to overcome the challenges of his brain injury and regain his independence.

The rehabilitation multidisciplinary team

The rehabilitation department at Quintomo Hospital is a state-of-the-art facility, with modern amenities and the latest in high-tech equipment. The walls are painted in soft, soothing colours and the rooms are spacious and well-lit, giving patients a sense of comfort and relaxation. Before the suspension of the humanoid robot program, robots were used to assist with tasks such as lifting and transferring patients, as well as providing emotional support.

Matthew Clark's rehabilitation team is made up of dedicated professionals who are experts in their respective fields. June Bailey, his main physiotherapist, is a gentle and encouraging presence, pushing him to achieve his goals while also understanding his limitations. Anju Sanjay, his main occupational therapist, is creative and innovative in finding ways to help him regain his independence and get back to the activities he loves. Mark Gower, his speech therapist, works with him to improve his communication skills and overcome any cognitive difficulties.

Other members of the team who worked with him during his rehabilitation program include Dr. Akira Nakagawa, the rehabilitation physician, and Dr. Brian Jordan, the rehabilitation medicine registrar. They monitored his progress and adjusted his treatment plan as needed, always keeping his goals and preferences in mind. The team also included psychologists, social workers, and dietitians, all working together to provide holistic care.

Matthew had fond memories of his time in rehabilitation and spoke highly of his team members. He remembered June's constant encouragement during physiotherapy sessions, Anju's creativity in finding ways to make everyday tasks easier, and Mark's patience during speech therapy sessions. He also appreciated the support he received from the other team members, who helped him navigate the emotional and psychological challenges of his recovery.

Despite the challenges he faced, Matthew made significant progress during his time in rehabilitation, regaining his strength and independence. The care and expertise of his rehabilitation team played a crucial role in his recovery, and he was grateful for their dedication and support.

Matthew Clark's discharge meeting

The multidisciplinary team assembled in the conference room at the rehabilitation department to discuss Matthew Clark's discharge plan. The atmosphere was relaxed, and everyone was in high spirits. The team had been working with Matthew for several weeks, and he had made impressive progress. They had grown fond of him and had become invested in his recovery.

June Bailey, his physiotherapist, began the meeting by presenting her assessment of Matthew's physical progress. She stated that he was now able to stand and walk independently, and his balance and coordination had significantly improved. Anju Sanjay, his occupational therapist, added that he had mastered his self-care tasks, such as dressing and grooming, and was ready to transition to home.

Mark Gower, his speech therapist, reported that Matthew's speech and language abilities had significantly improved, and he could now communicate effectively. The rest of the team members, including the rehabilitation physician, Dr. Akira Nakagawa, and the rehabilitation medicine registrar, Brian Jordan, all praised Matthew's remarkable recovery.

During the meeting, Matthew's wife, Sarah, and their two kids sat next to him, smiling and nodding along as the team discussed his progress. Dr. Michael Wong, the neurologist who had initially diagnosed Matthew's brain injury, also attended the meeting, along with Dr. Ryan Baker, intensivist, Dr. Amira Khan, interventional cardiologist, and Dr. Sally Brown, neurosurgeon. They all had cared for him early on during his admission.

Prof. Bill Stone, the Chief AI Officer, spoke about the research team's involvement in Matthew's care. He mentioned that they would check in over the phone monthly for the next year to monitor his progress and offer support and advice. The team nodded in agreement, pleased with the extra level of care.

As the meeting drew to a close, the team members congratulated Matthew on his remarkable progress and wished him well on his journey home. The doctors who cared for him during his admission also praised his resilience and determination, citing him as an inspiration to all.

Matthew Clark's Goodbye speech

Matthew Clark stood up to give his speech, his eyes filled with tears. He took a deep breath and began to speak.

"I can't believe I'm standing here today, speaking to all of you. Just a few weeks ago, I was lying in a hospital bed, fighting for my life. And now, here I am, standing on my own two feet, thanks to all of you."

He paused to wipe his tears and looked around the room at the team members who had saved his life.

"I want to express my gratitude to every single one of you. To the doctors who saved my life in the ICU, to the neurosurgeon who fixed my brain aneurysm, to Prof. Stone who recommended the clinical trial that saved my life, and to the rehabilitation team who helped me recover."

Matthew's voice started to crack as he continued.

"I know the hospital has been in the news for all the wrong reasons lately, but I want to say that my experience has been nothing but positive. The care I received here was top-notch, and I will be forever grateful to all of you."

He took a deep breath and continued.

"I don't think I could ever thank you enough for what you've done for me. But I hope you know that you've made a difference in my life, and I will never forget that."

Matthew's speech ended with a standing ovation from the audience. He then proceeded to hug each team member sitting in the front row, tears streaming down his face.

As the multidisciplinary meeting came to an end, the team dispersed, congratulating each other on a successful outcome. Prof. Bill Stone made his way over to Dr Wong, eager to hear an update on Dr Joshua Lee's engagement with the hospital psychologists and remediation process following detection of high blood alcohol levels by the sensors embedded in his scrubs. "Hey Michael, great work on your contribution to Matthew's progress. But how is Joshua doing? Has he been cooperative with the psychologists?" asked Prof. Stone.

Dr. Wong sighed heavily before replying, "I'm afraid not, Bill. Joshua has refused to engage with the psychologists and after a while, he stopped taking calls from the hospital. He was stood down from work while the investigation by the Australian Health Practitioner Regulation Agency (AHPRA) is underway."

Prof. Stone frowned at the news, "That's not good. Has there been any updates at all via any other channels?"

Dr Wong nodded, "Yes, actually. Today, the hospital lawyer received a demand letter from his lawyer."

Prof. Stone rubbed his forehead, "I've been in meetings all day and haven't had a chance to catch up on my emails. I'll look into it later this afternoon."

As they parted ways, Prof. Stone couldn't help but think that there may be more trouble on the horizon. "Oh my," he muttered to himself.

Chapter 19

THE 3CS WORKSHOP

As the attendees settled into their seats, Anita Bowles, the lead facilitator of the 3Cs workshop, introduced herself and her co-facilitators: Dr. Kirra Nakata, General Physician, David Mulvaney, Social Worker, and Matt Sutton, Communications Coach. She began the workshop by explaining the importance of the 3Cs: Compassion, Communication, and Cultural Competence in providing quality patient care. As Anita addressed the attendees, she emphasized the importance of the 3Cs workshop. "In this day and age, where technology is becoming increasingly integrated in healthcare, it's important that we don't forget the human element of care," she said. "The 3Cs - compassion, communication, and cultural competence - are the building blocks of patient-centred care, and that's what we're all here to provide. This is why Quintomo Hospital has made this workshop a mandatory annual training for all staff."

Some of the attendees nodded in agreement while others scribbled notes on their tablets. "This workshop is designed to help you hone these skills, so that you can provide the best care possible to our patients," Anita continued. "We want to make sure that our patients feel heard, understood, and respected, regardless of their backgrounds or circumstances."

As Anita wrapped up her introductory remarks, she added, "And just to let you all know, the 3Cs workshop is my personal favourite, but I also enjoy facilitating the wellbeing workshop, which focuses on the health and wellbeing of staff. After all, we need to take care of ourselves to be able to take care of others."

Anita led an exercise where attendees were paired up and asked to have a conversation with each other without using any verbal communication. "This is to illustrate how much we rely on nonverbal cues to communicate," she explained.

Matt, the communication coach, followed with a simulation where attendees were divided into groups and given different scenarios to role-play. "This exercise is to help you identify communication barriers that may arise when working with patients and their families," he said.

During the group session, Dr. Obioma Kalu and Dr. Sophie Bradley chatted about the importance of cultural competence in their respective fields. "As a urologist, I work with a diverse group of patients, so I need to be aware of how their cultural backgrounds and beliefs impact on their decisions," Dr. Kalu said.

Dr. Bradley, a resident in the emergency department, agreed. "It's important to understand how culture affects our patients' health decisions, and how we can communicate with them in such a way as to engender trust," she said.

Anita re-joined the group and continued with the workshop, emphasizing the importance of effective communication in building trust and rapport with patients. She gave examples of how miscommunication could lead to misunderstandings and how that could negatively impact patient outcomes.

The attendees were highly engaged and asked several questions, seeking clarification on certain aspects of communication, such as the use of medical jargon and how to handle difficult conversations with patients and their families.

As the workshop progressed, the attendees were deeply engaged in the discussion. David Mulvaney, the social worker, began a dialogue on providing compassionate care while maintaining professional boundaries. "It can be difficult to find the right balance," he said, "especially when we have patients who are going through tough times. But as healthcare providers, it's our responsibility to make sure we are not crossing any boundaries and to keep our patients' trust intact." The attendees chimed in, sharing their own experiences with the challenge of being compassionate while maintaining professionalism. "It's tough when we see our patients suffering and we want to do everything we can to help them," said Dr. Sophie Bradley, "but we have to remember to respect their privacy and maintain professional boundaries at all times." The attendees shared different strategies and tips on how to navigate this delicate balance, with some sharing personal stories of how they found it challenging at times.

During the workshop, Sophie, one of the resident doctors, shared her experience of working with an indigenous patient who spoke a dialect that was not available on Primeo. "It was tough using a lot of gestures and simplified language to explain the medical condition to the patient," she said. "But then, subsequently having an indigenous liaison officer present during the consultation made a world of difference. They were able to explain things more clearly to the patient and helped bridge the language gap."

Anita nodded in agreement, "Yes, having someone who can help with cultural and language interpretation can greatly improve communication and understanding between patients and healthcare providers."

Sophie chimed in, "It's not just about language barriers, but also understanding cultural differences and customs. It's important to be aware of our own biases and work towards providing culturally sensitive care."

Obioma, the new urologist, added, "I had a patient who refused treatment because they believed that traditional remedies were the only way to treat their condition. It made me realize that cultural beliefs can have a significant impact on how patients perceive and receive care."

Anita nodded in agreement and emphasized the importance of cultural competence in providing patient-centred care. "We need to recognize that patients come from diverse backgrounds with their own unique beliefs and values," she said. "Our role is to respect those differences and provide care that is tailored to their individual needs."

The attendees listened attentively and shared their own stories of working with patients from different cultural backgrounds. Some spoke of the challenges they faced, while others shared their successes in bridging cultural gaps and building rapport with their patients. The discussion was lively and engaging, and the facilitators were pleased with the level of participation and enthusiasm from the attendees.

During the morning tea break, a group of junior doctors huddled together, chatting amongst themselves. "Did you guys hear about Joshua Lee?" one of them asked. "He was stood down for working with high blood alcohol levels."

Another doctor shook his head. "That's not good. We need to be responsible and take care of ourselves. I'm glad we're having the wellbeing workshop later today."

The others nodded in agreement as they finished up their tea and headed back to the workshop room for the next session.

Overall, the 3Cs workshop was well-received by the attendees, who found the exercises and simulations engaging and informative. The workshop helped to reinforce the importance of compassion, communication, and cultural competence in the healthcare profession, and provided the attendees with practical tools to improve their interactions with patients.

The attendees left the workshop feeling energized and empowered, with a renewed commitment to providing patient-centred care that is compassionate, effective, and culturally sensitive.

Maria Santos, registered nurse, faints on duty

The strain on the hospital service had taken its toll on the nursing staff, who were already feeling the pressure from the suspension of the humanoid robot program. They were struggling to cope with the increased workload, and many were working long hours to try and keep up with the demand. Maria Santos, a 48-year-old registered nurse who was recently recruited from the Philippines, had been particularly hard-working, taking on extra shifts and putting in long hours to support her colleagues. However, the ongoing stress had begun to take a toll on her health.

One day, after using the toilet, she collapsed to the floor, landing heavily on her right side, and hitting her head.

The security team quickly responded to the emergency code triggered by Maria's sensors and broke into the bathroom to assess the situation. The medical team was also alerted and rushed to the scene. After confirming that Maria's vital signs were normal, they decided to transport her to the emergency department for further evaluation.

As they wheeled her towards the elevators, Maria slowly regained consciousness and asked, "What's going on?" The nurse, who was holding her hand, explained, "You passed out in the bathroom, but don't worry, you're in good hands now. We're taking you to the emergency department for further assessment and to make sure you're okay."

Maria looked confused and scared, but she trusted the nurse's reassuring words. "Am I going to be okay?" she asked, her voice trembling slightly. The nurse smiled at her and said, "We're doing everything we can to make sure you're okay. Just try to stay calm and breathe normally, okay?"

As they arrived at the emergency department, Maria was greeted by a team of doctors and nurses who quickly assessed her condition and ordered a CT scan of her brain. Maria tried to stay calm, but she couldn't shake the feeling of fear and uncertainty that had taken hold of her.

As she lay on the hospital bed, waiting for the results of her CT scan, Maria couldn't help but think about the strain on the nursing staff and the ongoing stress caused by the suspension of the humanoid robot program. She knew that her colleagues were doing their best to cope with the increased workload, but she also knew that they were human and prone to exhaustion and burnout.

Maria silently prayed for a positive outcome and hoped that she would soon be able to return to work and support her colleagues during these difficult times.

Dr. Jess Lewis, emergency physician, returned to Maria's bedside and gave her an update on her condition. "Your bloodwork and CT Brain came back normal," she said. "You suffered a vasovagal syncope secondary to dehydration and a concussion as a result of the head trauma."

Maria nodded, looking relieved to hear the news. "Will I be okay?" she asked.

"It's reassuring that you don't have any bleeding in your brain or other serious injuries.," Dr. Lewis explained. "But we would have to watch and wait to see how you recover from the impact to your head, "

Dr. Lewis then gave her some advice for her concussion precautions, "You'll need to take it easy for the next week, avoid strenuous activity, screen time and loud noises, and make sure to get plenty of rest. Also, avoid alcohol and medications that can affect your brain function."

Maria nodded and thanked the doctor for the advice. "So, can I go home now?" she asked.

Dr. Lewis smiled and said, "Yes, I'll give you a medical certificate for the week, and you can go home and rest. Just make sure to follow the concussion precautions and come back if you experience any headaches, nausea, or dizziness."

Maria thanked the doctor again and got ready to go home, relieved that her condition wasn't serious.

Director of nursing meets Maria's manager

Lakmini Abeysekera, the nurse unit manager for the neurology ward, had just finished reviewing the status reports on her staff fatigue levels when she received a video call from Jen Nguyen, the Director of Nursing. "Good afternoon, Jen," Lakmini greeted her, a hint of worry in her voice. "What brings you to call me on this busy day?"

"I received a report from Primeo about Maria Santos," Jen said, getting straight to the point. "How is she doing?"

"She's doing alright," Lakmini said, relieved to hear that the call was about Maria's condition and not some new crisis. "Her bloodwork and CT scan were normal, and she's been advised to take the week off due to the concussion."

"That's good to hear," Jen said, but Lakmini could see that she wasn't fully convinced. "I've been reviewing some of the reports on staff fatigue levels, and it seems like the numbers are higher than usual. Can you give me a rundown of the process for flagging fatigue levels and how you handle each case?"

Lakmini nodded. "Usually, after sensors on staff's scrubs detect trends suggestive of fatigue they are flagged as potentially fatigued. Then Primeo sends an automated self-check to the staff member, asking them to rate their fatigue level on a scale of 1 to 10. Next they are asked if they feel strong and capable of continuing to work, if they need to take an immediate break, or if they need to head home immediately because they are impaired."

Jen nodded along as Lakmini spoke. "And how do you handle each case once they're flagged?"

"We review each case individually," Lakmini explained. "One-third of the cases reassure us that they're fine and usually don't run into problems. Another one-third need a break to rest and can complete their shift. And the last one-third voluntarily decide to go home after self-check."

Jen looked concerned. "And how has that process been working since the suspension of the humanoid program?"

Lakmini sighed. "The levels of fatigue flagged have been up to 50% of staff from around 10% previously, and we're having difficulty reviewing each individual case. It's becoming overwhelming."

She added, "I'm sorry Jen, I didn't get the chance to review Maria's fatigue alert, the workload has been overwhelming for me and my team. But I have since learnt that she completed her automated self-check earlier today. She indicated that she needed to take a break and was waiting for another nurse to return from her afternoon tea break before taking hers."

Jen nodded understandingly, "Yes, I see that here on her status report. It's a shame that the staffing constraints have made it difficult to provide adequate support for our staff during these challenging times."

Lakmini nodded in agreement, "It's been a struggle, but we are doing our best to manage the situation. We have implemented some measures to help alleviate the workload, such as shifting some responsibilities to the support staff and encouraging staff to take breaks when needed."

Jen leaned in, "I understand the strain on your team, Lakmini. We are working on a plan to address the staffing constraints and hope to have some solutions soon. In the meantime, please ensure that your team members are taking care of themselves and taking the necessary breaks to prevent burnout."

Lakmini nodded, "Thank you, Jen. I will make sure to remind my team of the importance of self-care and taking breaks when needed."

Jen Nguyen prepares for her presentation to the board

Jen sat at her desk, staring at the report in front of her. "Primeo, I need your help," she said aloud. The AI assistant beeped to indicate it was listening. "I need a title for this report that summarizes the impact of the humanoid robot program suspension on our nursing staff."

"Sure, Jen. Here are three suggested titles: 'Navigating Nursing Staff Fatigue in the Absence of Humanoid Robots Assistance,' 'Overcoming the Challenges of Staffing and Care in a Post-Humanoid World,' and 'Redefining Patient-Centered Care Amidst Technology Limitations.' Which one do you prefer?"

Jen considered the options. "I think 'Navigating Nursing Staff Fatigue in the Absence of Humanoid Robots Assistance' is the best fit. Let's go with that one."

"Great choice, Jen. I have updated the report title accordingly."

Jen nodded, satisfied with the choice. "Thank you, Primeo. Can you also go through the report and suggest any changes or additions that need to be made?"

"Of course, Jen. I have already gone through the report and made some suggestions. I suggest we add some statistics on incidents of staff errors, fatigue alerts, and patient complaints to reinforce the impact of the program suspension. Additionally, we can include some recommendations for how to address these challenges moving forward."

Jen nodded again. "That's a good idea. Let's make those changes."

"Understood, Jen. I will make the revisions and update the report accordingly." Jen continued to reflect and prepare for her presentation to the board of directors of Quintomo Hospital due the next day.

Chapter 20

Board Meeting in Crisis: Quintomo Hospital Faces Tough Decisions

Amid the crisis, the Quintomo Hospital leadership team convened a board meeting to brainstorm solutions and update key stakeholders on the hospital's performance and progress. The meeting was crucial to the hospital's success and attended by board members and the executive team. At the forefront of the meeting was the Director of Nursing, Jen Nguyen, who was eager to push for the speedy reinstatement of the humanoid robot program.

Dr Abdul Sheikh, Executive Director of Medical Services at Quintomo Hospital, gave a presentation on how the hospital's doctors have adapted to the new challenges that have arisen since the suspension of the humanoid robot program. Being one of the leading hospitals in AI and health technology in the world, Dr Sheikh emphasized that the doctors have had to show patience, sensitivity, and understanding in dealing with the increased demands on the nurses and patient support services. He also noted that doctors have not been directly impacted since they do not do patient handling, but they have had to adapt to the new environment and management plans by improving their communication and collaboration skills. One of the ways that doctors have been affected is having to perform CPR manually during resuscitation, as they had to step up to fill the gap left by the absence of the humanoid robots. Sheikh also touched on the importance of regular training and updating of emergency protocols to ensure that the doctors are prepared for any situation that may arise in the future.

The Director of nursing makes her case for the reinstatement of the humanoid robots

Jen Nguyen, the Director of Nursing, was next to present her report to the board of directors at Quintomo Hospital. Her presentation was titled 'Navigating Nursing Staff Fatigue in the Absence of Humanoid Robots Assistance.' As she began her presentation, Jen acknowledged the difficult year the hospital had experienced, from the cyberattack to the suspension of the humanoid robot program, but she also emphasized the resilience and determination of the hospital staff.

Jen highlighted the critical role that technology played in improving patient outcomes and quality of care. She noted how their work was made efficient with the aid of Primeo, the hospital's integrated AI assistant, communications system, and electronic medical records system. "Primeo has streamlined documentation and improved communication among healthcare providers", she said. However, she also stressed the importance of balancing technology with the human touch, ensuring that patients received compassionate, patient-centred care.

Jen then went on to discuss the impact of the suspension of the humanoid robot program on the nursing staff, noting the increased workload and stress levels among nurses. She emphasized the urgent need to reinstate the program, highlighting the benefits of the program in improving patient safety, reducing the risk of infections, and reducing the workload on nurses. Jen shared data from patient feedback surveys and staff fatigue alerts, which showed a clear need for the program's reinstatement.

In her presentation, Jen also outlined her plans, including expanding the hospital's telemedicine services and investing in further staff training and development. She emphasized the importance of staying at the forefront of technological advancements in healthcare and continuously improving the patient experience.

As she concluded her presentation, Jen thanked the board for their continued support and urged them to prioritize the reinstatement of the humanoid robot program to improve patient care and support the nursing staff.

The board responds

During the board meeting, Geoff Skye, the chairman of the board, asked about Simon, the registered nurse who was punched in the face by a confused and agitated patient. "How is Simon doing?" he asked.

Jen Nguyen, the Director of Nursing, responded, "Simon is doing well, thank you for asking. He's currently on leave and receiving support from our Employee Assistance Program.

We're also implementing new safety measures to prevent similar incidents from happening in the future."

Prof. Bill Stone, the Chief AI Officer, then enquired about Maria Santos, the nurse who collapsed in the bathroom. "How is Maria doing?" he asked.

Jen Nguyen responded, "Maria is doing much better, thank you for asking. She's following her doctor's orders and taking some time off to rest and recover."

Other board members expressed their concern and support for both Simon and Maria, acknowledging that the incidents could have had much worse outcomes. They also commended the staff for their handling of the situations and for supporting each other through the difficult times.

The boardroom was tense as the topic of the reinstatement of the humanoid robot program was brought up for discussion. Jen Nguyen, the Director of Nursing, argued for immediate reinstatement citing their internal review which showed their systems were not compromised and the security upgrades that have already been carried out.

Geoff Skye, the Chairman of the board, was concerned and asked, "What about the reports from CyberPower and the Therapeutic Goods Administration (TGA)? We cannot risk the safety and security of our patients and staff."

Prof. Bill Stone, the Chief AI Officer, added, "We need to take our time and review the reports before we make any hasty decisions. The last thing we want is a repeat of the cyberattack."

Jen retorted, "But our patients and staff are suffering without the support of the humanoid robots. We cannot wait indefinitely for those reports."

Patrick Macguire, the CEO of Quintomo Hospital, chimed in, "Jen has a point. Our staff is already stretched thin, and the longer we wait, the worse it gets."

A board member, Ahmad, argued, "But what about the potential risks? Can we afford to take that chance?"

Another board member, Leonie, added, "And what about the public perception? The media will have a field day if we make the wrong decision."

The debate went back and forth for some time, with each member presenting their arguments and concerns. Emotions ran high, and at times the arguments became heated.

Finally, after much discussion, it was agreed that the hospital would wait for the reports from CyberPower and the TGA before making a final decision on the reinstatement of the humanoid robot program. However, in the meantime, they would explore other options for easing the strain on their staff, including hiring additional support staff and increasing training and support for their current staff.

The board members had a sense of relief that they had reached a decision, albeit a temporary one. The safety and well-being of their patients and staff were paramount, and they could not afford to take any chances.

Legal matters

Andy Sumner, Director of legal services at Quintomo Hospital, began his presentation by highlighting the importance of legal compliance in healthcare institutions. He provided an overview of the current legal issues facing the hospital, including medical malpractice claims, employee disputes, and data privacy breaches.

Sumner went on to discuss the various ways in which the hospital is managing these legal issues. He presented some cases to illustrate each of the five talking points he had prepared.

The first case involved a patient who had suffered complications following a routine surgery. The patient filed a medical malpractice claim against the hospital, alleging negligence on the part of the surgeon and hospital staff. Sumner explained that the hospital had responded by conducting a thorough investigation into the incident, working with external legal advisors and the insurance company to reach a fair settlement with the patient.

In the second, Sumner discussed an employee dispute that had arisen between a senior nurse and a junior nurse. The senior nurse had accused the junior nurse of misconduct, which had resulted in a formal complaint being filed. The hospital had responded by investigating the matter and mediating a resolution between the two parties, which allowed them to continue working together.

The third case study focused on a data privacy breach that had occurred when an employee accidentally leaked patient data. Sumner outlined how the hospital had responded by taking immediate action to contain the breach, notifying the affected patients, and implementing new policies and training programs to prevent similar incidents from happening in the future.

In the fourth case, Sumner discussed a dispute that had arisen between the hospital and a supplier over a contract. He explained how the hospital had worked with external legal advisors to negotiate a resolution that was beneficial to both parties.

Finally, in the fifth case, Sumner discussed a case involving a nurse who had been accused of stealing medication from the hospital. The nurse had been fired and had filed an unfair dismissal claim against the hospital. Sumner explained how the hospital had worked with external legal advisors to successfully defend the claim and ensure that the nurse did not return to work at the hospital.

Throughout his presentation, Sumner emphasized the importance of legal compliance and risk management in healthcare institutions. He urged the board to continue supporting the hospital's legal team and to prioritize legal compliance in all areas of the hospital's operations.

The board members responded with thoughtful questions and comments, acknowledging the importance of legal compliance in ensuring the hospital's success. They expressed their gratitude to Sumner and the legal team for their hard work and dedication to managing the hospital's legal issues.

After the presentation, the board took a short break to review the information presented before moving on to the next item on the agenda.

The board discuss Dr. Joshua Lee impending legal case

Andy Sumner cleared his throat and began, "As you all may be aware, we have received a demand letter from the lawyers representing Joshua Lee, the first-year neurology trainee who was stood down following detection of high blood alcohol levels by his sensors."

The room fell silent as Sumner continued, "Dr Lee's lawyers claim that the hospital's monitoring and tracking practices are an invasion of privacy and infringe on his rights. They have threatened to take legal action if he is not reinstated and compensated for the damage caused."

A murmur of concern swept across the room as some members of the board exchanged worried glances.

Sumner took a deep breath before continuing, "We understand the gravity of the situation and will be beginning preliminary discussions with Dr. Lee's lawyers tomorrow. However, I must point out that our monitoring and tracking practices are in place to ensure the safety and well-being of our staff and patients. These measures have been put in place to reduce errors, prevent accidents, and ultimately save lives."

Prof. Bill Stone, the Chief AI Officer, interjected, "Our AI technology is constantly evolving and improving, and we are always mindful of the need to balance privacy with safety concerns. The sensors and smart speakers in use at Quintomo Hospital are designed to help us provide the best possible care to our patients while ensuring the safety and well-being of our staff."

Jen Nguyen, the Director of Nursing, added, "I think it's important to emphasize that our staff have a responsibility to ensure they are fit for duty when they come to work. Joshua's situation is regrettable, but our policies are in place to protect both our staff and patients."

The board members exchanged nods of agreement as Sumner continued, "Furthermore, our policies and practices are in compliance with all relevant laws and regulations, including those pertaining to data privacy and confidentiality. We have taken every measure to ensure that staff members are fully aware of their rights and obligations regarding monitoring and tracking, and that they have the ability to raise their objections over any features they are uncomfortable with."

Sue, one of the board members, spoke up, "I understand the need for safety measures, but we must also consider the privacy concerns of our staff. Are there any steps we can take to strike a better balance between privacy and safety?"

Sumner responded, "We are always open to suggestions and feedback from our staff and stakeholders, and we take all concerns seriously. We are currently reviewing our policies and practices to ensure that we are striking an appropriate balance between privacy and safety."

As the meeting continued, the board members discussed the issue in detail, with some expressing concern about the potential legal implications of the case. Others offered suggestions for how the hospital could better educate and inform its staff about its monitoring and tracking policies.

Overall, the discussion highlighted the need for Quintomo Hospital to continue to strike a balance between privacy and safety concerns, while ensuring that its policies and practices remain compliant with all relevant laws and regulations.

Chapter 21

THE VERDICT: AHPRA'S DECISION ON JOSHUA'S CASE

Dr. Joshua Lee was at his childhood home in Toowoomba, Queensland, where he received support from his parents, Dr. David Lee and Mrs. Maria Lee. His father is a retired pathologist, and his mother a practicing lawyer, although not involved in his case. Despite the allegations against him, Joshua maintains his innocence and is determined to clear his name. Over dinner, the family discussed the case, mapping out a strategy and debating the best way to proceed. The atmosphere was tense as they weighed the potential consequences of the charges against Joshua, but his parents remained resolute in their support of their son.

As Joshua sat down to dinner with his parents in their cozy home in East Toowoomba, they exchanged small talk about their day, discussing the latest news and updates. Despite the serious nature of Joshua's situation, they tried to keep the conversation light-hearted and jovial.

"So, how was your day dad?" Joshua asked his father, who had retired from his position as a pathologist at a nearby hospital a few years ago.

"It was good, thanks for asking. Just did some gardening and caught up on my reading," his father replied with a smile.

"And how about you, mum? Any exciting cases today?" Joshua asked his mother, who was a practicing lawyer in the area.

"Well, nothing too exciting, but I did have a few interesting consultations with some new clients," his mother replied, her tone light and relaxed.

After a few more minutes of small talk, the conversation turned to the reason why they were all gathered: Joshua's case. They discussed the latest updates, including the verdict from the Australian Health Practitioner Regulation Agency (AHPRA) that was released that day. AHPRA mandated Joshua to undertake counselling for 3 months and upon satisfactory completion, he could return to work with ongoing extra supervision and psychologist support for another 6 months.

"I still can't believe they're making you go through all of this," Joshua's father said, shaking his head in disbelief.

"I know, dad, but I didn't do anything wrong. I maintain my innocence, and I'll fight this to the end," Joshua replied, determination in his voice.

His mother chimed in, "We're here for you, Joshua. We'll do whatever it takes to make sure you get the justice you deserve."

Joshua's stance on his AHPRA disciplinary recommendations

Joshua's parents expressed relief that he could return to work after completing the mandatory counselling sessions. They discussed the conditions of his return, including the additional supervision and support he would receive for the next six months. But their relief was short-lived when Joshua vehemently refused to attend any counselling sessions as recommended by AHPRA. "I'm innocent, and I don't need counselling," Joshua declared. "I didn't drink any alcohol before work, and I certainly don't usually drink enough to be impaired. Quintomo Hospital violated my privacy rights by using those sensors to monitor my blood alcohol levels without my consent."

His parents tried to reason with him, but Joshua was adamant about his decision. "I'll see Quintomo Hospital in court," he said. "I want to clear my name and hold them accountable for what they've done."

They discussed the potential outcomes of suing the hospital and the challenges that could come with it. Joshua's mother, a lawyer, cautioned that the process could be lengthy and emotionally draining.

Despite his parents' concerns, Joshua remained determined to take legal action against the hospital. He believed it was the only way to clear his name and expose the hospital's questionable monitoring practices.

Joshua sat at the dinner table with his parents, looking down at his plate as they continued to strategize on how to get justice for him. After a few moments of silence, he cleared his throat and said, "I've hired a law firm to represent me in this matter."

His parents looked up in surprise. "Who have you hired?" his father asked.

Joshua took a deep breath. "I've hired the law firm of Pearson, Robertson, and Associates. They specialize in healthcare law and have a lot of experience with cases like mine."

His mother nodded in approval. "That sounds like a good choice, dear. Do you have a meeting with them soon?"

"Yes, tomorrow morning. And they've invited representatives from Quintomo Hospital to join us for a preliminary meeting."

His father raised an eyebrow. "Who will be representing you from the law firm?"

"There will be two senior lawyers," Joshua replied. "The lead lawyer will be Michael Pearson. He's a highly respected healthcare lawyer with over 20 years of experience. He's been involved in some high-profile cases and has a reputation for being tough but fair."

"And who will be assisting him?" his mother asked.

"The other lawyer will be Sarah Robertson," Joshua said. "She's also a healthcare lawyer, but with a focus on privacy and data protection. I think she'll be able to help me argue that the hospital violated my privacy rights by monitoring me without my knowledge or consent."

His parents nodded thoughtfully. "It sounds like you've got a good team behind you," his father said.

Joshua nodded, feeling more confident now that he had legal representation. "I just want to clear my name and prove that I didn't do anything wrong. And if the hospital violated my privacy, then they need to be held accountable."

His parents gave him a reassuring smile, telling him they were behind him all the way.

Joshua Lee and his parents strategize

As they mapped out their strategy, the discussion focused on the hospital's monitoring and tracking practices, with Joshua arguing that the hospital's policies went too far in infringing on employees' privacy. "I never signed up for a job where I'm constantly being monitored and tracked," he said. "They have no right to monitor my every move while I'm at work."

His mother acknowledged the potential weight of this argument, but pointed out that the hospital may argue that such monitoring is necessary for the safety of patients and employees. "We need to be prepared for the hospital to argue that they have a responsibility to keep everyone safe, and that tracking employees' movements and health is part of that responsibility," she said.

Joshua also planned to argue that the hospital's policies were discriminatory, as they only applied to certain employees and not others. "If they're going to track and monitor us, they should be doing it for everyone, not just the doctors and nurses," he said.

His mother agreed that this was an important point but cautioned that the hospital may argue that tracking nurses and doctors was necessary due to the nature of their work. "We need to be prepared for the hospital to argue that tracking healthcare workers is necessary to prevent the spread of infection and ensure patient safety," she said.

As they continued to discuss the case, they also talked about the potential consequences of Joshua's decision to sue the hospital. His father expressed concern over the financial and emotional costs of a lengthy court battle, but Joshua was resolute. "I maintain my innocence, and I won't let the hospital's invasion of my privacy go unpunished," he said.

With their discussion points mapped out, Joshua prepared for his meeting with the lawyers the next day. They all agreed that the next few months would be challenging, but they were ready to face them together as a family.

Joshua receives a call from Mo

Joshua and his parents finished dinner, expressing hope that the ordeal would be over soon. After they dispersed from the dinner table, Joshua received a phone call from his partner, Mo.

"Hey babe, how was your day?" Joshua asked as he answered the phone.

"It was okay, just busy with work," replied Mo, who works as a prompt engineer at a Quintomo-based Biotech startup.

"I hear you. How's everything going with you?" Joshua asked.

"I'm doing alright. Just worried about you, you know," Mo said.

"I appreciate it, but I'm holding up okay. Just trying to stay positive," Joshua replied.

"I know, but it's not fair what they're doing to you. I can't believe they're making you go through all this," Mo said.

"I know, it's been tough, but I have some good news. The preliminary meeting between my lawyers and the hospital is scheduled for tomorrow," Joshua said.

"That's really good. Do you feel optimistic about it?" Mo asked.

"I do. My lawyers have a great track record, and I feel like they really understand my situation," Joshua said.

"Well, that's a relief. I'm glad you have someone on your side," Mo said.

"Me too. Thanks for being there for me, Mo. It means a lot," Joshua said.

"Of course, I'm here for you always. We'll get through this together," Mo replied.

With that, they chatted a bit more about their day before saying their goodnights and hanging up the phone. Joshua felt comforted by Mo's support and was grateful for her unwavering love and encouragement.

As Joshua lay in bed, he couldn't shake off the unease he felt about the meeting with the lawyers and the hospital the next day. He knew that the road ahead would be long and difficult, but he was determined to fight for his rights and clear his name. The thought of being accused of something he believed he didn't do weighed heavily on him, and he couldn't help but wonder how his life would turn out if he lost the case.

As he tried to push these thoughts aside and drift off to sleep, he couldn't help but think about Mo and how much she had been there for him throughout this ordeal. He felt grateful for her support and the love they shared.

Despite his anxiety, Joshua knew that he had to stay strong and focused on the task ahead. He vowed to himself that he would fight with everything he had, knowing that his reputation, career, and future were at stake.

As he closed his eyes, Joshua took a deep breath and tried to clear his mind, hoping to get some rest before the meeting the next day.

Chapter 22

THE BATTLE BEGINS: JOSHUA'S DEFENCE TAKES SHAPE

Joshua sat anxiously in the conference room of Pearson, Robertson and associates. His lawyers, Michael Pearson and Sarah Robertson, sat across from him. The room was adorned with shelves lined with law books and legal memorabilia, creating an atmosphere that made Joshua feel both nervous and determined.

Michael broke the silence, his voice calm and reassuring. "Joshua, let's go over the details of what transpired at Quintomo Hospital. It's important for us to have a clear understanding of the events leading up to your suspension."

Joshua took a deep breath, his mind racing with the memories of that fateful day. "I had just finished my shift in the neurology ward. It had been a particularly stressful day with multiple urgent cases coming in. When I finally got off work, I went to a nearby bar with some colleagues to unwind. I admit, I had a few drinks to relax, but I never thought it would lead to this."

Sarah leaned forward, her eyes filled with empathy. "Joshua, can you recall if anything happened at the bar that could have affected your blood alcohol level? Were there any unusual circumstances or encounters that you remember?"

Joshua furrowed his brow, his mind searching for any significant details. "Honestly, nothing out of the ordinary happened. We were simply enjoying our time together, and I didn't feel intoxicated. It's possible that the sensors malfunctioned or there was some other error in the system."

Michael nodded, taking notes as Joshua spoke. "We'll thoroughly investigate the accuracy of the sensors and review the hospital's protocols. It's crucial to establish whether there were any discrepancies or technical issues that may have led to an incorrect reading."

Sarah leaned back in her chair, her expression thoughtful. "Joshua, it's important for us to understand your perspective and the impact this suspension has had on your life. Can you share with us how this incident has affected you personally and professionally?"

Joshua's gaze dropped for a moment, his voice tinged with frustration. "This suspension has tarnished my reputation and cast doubt on my abilities as a doctor. It's been incredibly challenging to face the judgment and speculation from my colleagues and the wider medical community. I've worked hard to get to where I am, and now it feels like everything is crumbling."

Michael leaned forward, his tone firm but supportive. "Joshua, we understand the gravity of this situation, and we are here to fight for your rights. We will leave no stone unturned in seeking the truth and ensuring that your reputation is restored."

Sarah leaned forward, her eyes focused intently on Joshua. She took a deep breath, ready to delve into the details that could potentially unravel the mystery surrounding his high blood alcohol levels.

"Joshua, let's go back to the night before your sensors detected elevated alcohol levels," Sarah began, her voice steady and probing. "You mentioned having a few beers at the bar, but I need to understand the specifics. What type of beer were you drinking?"

Joshua furrowed his brow, his memory straining to recall the exact details. "It was a local craft beer, I believe. I don't recall the brand specifically."

Sarah nodded, jotting down notes. "And aside from beer, did you consume any other types of alcohol that evening? Wine or spirits, perhaps?"

Joshua hesitated for a moment, his mind replaying the scenes from that night. "No, just beer. I didn't have any wine or spirits."

Sarah pressed further, her voice unwavering. "Can you remember the name of the bar where you and your colleagues gathered?"

Joshua racked his brain, trying to summon the name from the depths of his memory. "It was called 'The Oak Barrel'. It's a popular spot among hospital staff."

Sarah's eyes narrowed slightly, sensing there might be more to uncover. "Who were the colleagues with you at the bar that evening?"

Joshua paused, recalling the faces and names of those who had joined him. "There were a few of us from the neurology department—Dr. Rachel Thompson, Dr. Andrew Patel, and Nurse Emma Clarke. We often go out together to unwind after our shifts."

Sarah noted down the names, her mind piecing together the puzzle. "And what time did you leave the bar that night, Joshua?"

Joshua took a moment to calculate the time. "I believe it was around 10:30 pm. I didn't want to stay too late as I had an early shift the next morning."

Sarah's tone became more insistent as she continued her line of questioning. "Once you left the bar, did you consume any more alcohol before going to bed?"

Joshua's eyes widened as he realized the significance of the question. "No, absolutely not. After leaving the bar, I went straight home and went to bed. I didn't have any more alcohol."

Michael Pearson interjected, his voice measured and calculated. "Joshua, we need to establish a clear timeline of events and gather any supporting evidence. It's important that we ascertain the facts surrounding that evening to build a strong case in your defence."

Joshua nodded, his determination growing. "I understand, and I'm confident that the truth will prevail. I'm innocent, and we will find the evidence to support that."

As Sarah continued her meticulous questioning, the room grew heavy with anticipation. Every detail, no matter how seemingly insignificant, held the potential to shape Joshua's defence and uncover the truth. The stage was set for a fierce legal battle, as the pieces of the puzzle slowly started to fall into place.

The hospital's legal team meet with Joshua and his legal team

The meeting between Joshua and his lawyers and the hospital's lawyers was set in a conference room in the city. Joshua was accompanied by his lawyers, Michael Pearson and Sarah Robertson, while the hospital's legal team was headed by Andy Sumner, the Director of Legal Services. They all took their seats around a large conference table, with Joshua seated between his lawyers.

The case involved Dr. Joshua Lee, a first-year neurology advanced trainee at Quintomo Hospital who was stood down after his sensors detected high blood alcohol levels. Joshua had maintained his innocence and refused to attend counselling sessions as recommended by the Australian Health Practitioner Regulation Agency (AHPRA). He believed that Quintomo Hospital had violated his privacy rights and was planning to sue them. Joshua had engaged the services of a well-known law firm, and they were all present at the meeting, ready to discuss the case.

Andy Sumner began the meeting by outlining the hospital's position. He explained that the hospital had taken the necessary steps to ensure the safety and well-being of its staff and patients, which included the implementation of the monitoring and tracking systems. He noted that the hospital had a responsibility to ensure that its staff were fit to work, and that patient care was not compromised.

Sarah Robertson asked for more details about the hospital's monitoring and tracking systems, and the data they collected. Sumner explained that the systems were designed to monitor staff fatigue, location, and health status. He also noted that the data collected was used to ensure that the hospital was providing the best care possible and to identify areas for improvement.

Joshua's lawyers expressed concern that the hospital's monitoring and tracking systems may infringe on staff privacy rights. Sumner responded by explaining that the systems were designed with privacy in mind, and that the hospital had taken measures to protect staff privacy rights. He noted that staff were informed about the systems and that they were only used for specific purposes related to staff safety and patient care.

Michael Pearson questioned the reliability of the monitoring and tracking systems, particularly in detecting alcohol levels. Sumner responded by stating that the hospital's systems were tested and calibrated regularly to ensure their accuracy. He also noted that the systems had been used effectively in the past to detect staff impairment, which had helped prevent potential safety incidents.

The hospital's legal team expressed their willingness to resolve the matter through mediation, rather than proceeding to court. They noted that court proceedings could be costly and time-consuming, and that mediation could provide a more expedient and mutually beneficial outcome.

Joshua's lawyers began their presentation by outlining the events leading up to Joshua's suspension, including the monitoring and tracking of staff using sensors embedded in their scrubs. They argued that the hospital's monitoring practices had violated Joshua's privacy and that the decision to stand him down was unjustified.

"Mr. Lee has maintained his innocence throughout this entire ordeal," Joshua's lead lawyer, Michael Pearson, stated. "We believe that the hospital's actions were unwarranted and unlawful, and we intend to pursue this matter to the fullest extent of the law."

He then went on to discuss AHPRA's disciplinary measures and the impact of the false accusation on Joshua's career and reputation.

"We have reason to believe that the hospital's monitoring practices extend beyond what is necessary to ensure patient safety," Michael Pearson continued. "In fact, we believe that they may be in violation of both Australian law and international human rights standards.", Sarah Robertson, Joshua's other lawyer added.

The hospital's lawyers responded to Joshua's lawyers' claims by presenting their own evidence, including the hospital's internal policies and procedures, and the steps they had taken to ensure staff privacy was protected.

"We take our responsibility to our staff and patients very seriously," one of the hospital's lawyers, Mark Johnson, stated. "While we understand Dr. Lee's concerns, we believe that our monitoring practices are necessary to ensure the safety and well-being of our patients."

The discussion then moved towards the possibility of mediation, with Mark Johnson, expressing their willingness to resolve the matter through this method. However, Sarah Robertson quickly shut down any notion of mediation.

"Dr. Lee has suffered greatly as a result of the hospital's actions, and we believe that the only way to rectify this is through legal action," she stated firmly. "We are prepared to take this matter to court if necessary."

The hospital lawyers looked at each other and then turned to Michael Pearson, Joshua Lee's lead lawyer. Sarah Robertson, Joshua's other lawyer, continued speaking:

"Nevertheless, we would like to explore the possibility of resolving this matter through mediation. We believe that this would be the best course of action to avoid prolonged litigation and come to a mutually agreeable resolution."

The hospital lawyers looked at each other again before responding.

"We are willing to consider mediation as an option, but we believe that it's important to fully investigate all aspects of this case before coming to any conclusions."

Joshua looked at his lawyers, frustrated with the hospital's response. "I'm sorry, but I can't agree to mediation. I believe that I have been wronged and I want my day in court."

Michael and Sarah exchanged a look before turning back to Joshua.

"We understand your position, Joshua. We will do everything in our power to ensure that your rights are protected, and that justice is served."

The meeting continued for a while longer, with both sides discussing their perspectives and presenting evidence to support their claims. Joshua's lawyers continued to press the hospital's lawyers on their monitoring and tracking practices, while the hospital's lawyers reiterated their commitment to patient safety and quality of care.

As the meeting ended, both sides agreed to continue to explore all possible avenues for resolution, including mediation. At the next meeting, Joshua Lee's team would be presenting their terms for mediation.

Chapter 23

THE HUMAN TOUCH OF ROBOTICS: DR. OBIOMA KALU'S APPROACH TO COMPASSIONATE UROLOGICAL CARE

Dr. Obioma Kalu, known to his colleagues as Obi, was a highly esteemed urologist who had recently joined the team at Quintomo Hospital. Originally from Nigeria, he had made a name for himself during his time at Harvard Medical School, where he completed his training and pursued his passion for robotic surgery. With a strong background in urological research, Dr. Kalu had also earned his Ph.D. by focusing on the application of artificial intelligence in predicting the risk of urological cancers.

Despite the unfortunate setback of the recent cyberattack that had rocked Quintomo Hospital shortly after his arrival,

Dr. Kalu remained steadfast in his belief in the power of AI and robotics to revolutionize urology. He was optimistic about the potential of these advanced technologies to improve patient outcomes, enhance surgical precision, and streamline diagnosis and treatment processes.

Dr. Kalu quickly established himself as a dedicated and passionate member of the hospital's urology department. He brought with him a wealth of knowledge and expertise, and his colleagues respected him for his innovative thinking and commitment to advancing the field. Dr. Kalu was not only focused on providing exceptional patient care but also dedicated to conducting groundbreaking research in the realm of robotic surgery and AI applications in urology.

In his free time, Dr. Kalu spent time with his young family, and immersed himself in the vibrant culture and diverse community of Quintomo. He appreciated the warm welcome he had received from his colleagues and the hospital staff, which helped him feel at home despite being far away from his native country. Dr. Kalu looked forward to making significant contributions to the field of urology at Quintomo Hospital, collaborating with his colleagues, and embracing the opportunities for research and innovation that lay ahead.

As he prepared for the day's work, Dr. Kalu's excitement and determination were palpable. With his expertise in robotics and AI, he was eager to explore new frontiers, push the boundaries of urological care, and continue his pursuit of excellence in patient outcomes. With a resilient spirit and a burning passion for his work, Dr. Kalu embarked on each day at Quintomo Hospital with a sense of purpose, ready to make a lasting impact on the future of urology.

Dr. Kalu's work setting

Dr. Obioma Kalu's clinic room in the urology department of Quintomo Hospital was a testament to the advancements of modern technology in healthcare. The room was designed to optimize efficiency and enhance patient care through the integration of advanced equipment and digital tools. As Dr. Kalu entered his clinic room, he was greeted by a state-of-the-art environment that enabled him to provide the highest level of care to his patients.

The centrepiece of the room was a series of multi monitors strategically positioned to provide Dr. Kalu with a comprehensive view of patient information, medical records, and imaging data. These high-resolution displays allowed him to review and analyse complex images and diagnostic reports with remarkable clarity and precision. With a simple touch, Dr. Kalu could navigate through various imaging modalities, seamlessly comparing scans and zooming in on specific areas of interest.

To further enhance his diagnostic capabilities and surgical planning, Dr. Kalu utilized augmented reality (AR) and virtual reality (VR) technologies. Through specialized headsets and software applications, he could visualize three-dimensional reconstructions of patients' anatomical structures. This immersive experience provided him with invaluable insights and aided in accurately mapping out surgical procedures and treatment plans.

In addition to the advanced imaging and visualization technologies, Dr. Kalu had access to telemedicine capabilities within his clinic room. This allowed him to remotely connect with patients, colleagues, and specialists from around the world for virtual consultations and collaborative discussions. Through secure video conferencing platforms, he could review cases, share medical images, and provide expert opinions, ensuring that patients received the best possible care regardless of their location.

Dr. Kalu's clinic room was thoughtfully designed to maximize efficiency and streamline patient care. The room also featured comfortable seating areas for patients and their families, fostering a welcoming and compassionate environment.

Adjacent to his clinic room was the urology procedure suite, where Dr. Kalu performed a variety of minimally invasive procedures and interventions. Equipped with advanced endoscopic tools and instruments, this specialized suite allowed him to conduct cystoscopies, perform biopsies, and carry out other urological procedures with precision and minimal invasiveness. The suite was staffed by a dedicated team of highly trained nurses and anaesthetists who worked collaboratively with Dr. Kalu to ensure the highest level of patient safety and comfort.

While Dr. Kalu's clinic room and the urology procedure suite were situated on the same level of the hospital, the urology theatres were located in another section of the building. These state-of-the-art operating rooms were equipped with cutting-edge robotic systems that Dr. Kalu utilized for complex surgeries. The robotic platforms enabled him to perform intricate procedures with enhanced precision, control, and dexterity, resulting in shorter recovery times and improved patient outcomes.

Dr. Kalu relied on Primeo, the hospital's integrated AI assistant, to streamline his workflow and access critical patient information. Primeo Glass, a wearable device, allowed him to review multiple high-definition images and data on a virtual workstation, eliminating the need for physical monitors and facilitating seamless interaction with patient records. This innovative technology enhanced Dr. Kalu's diagnostic capabilities and supported efficient decision-making, ultimately benefiting his patients.

In this advanced and technology-driven work setting, Dr. Obioma Kalu embraced the opportunities presented by AI, robotics, and telemedicine. With his dedication, expertise, and the support of cutting-edge tools, he aimed to provide the highest standard of care to his patients and push the boundaries of urological medicine at Quintomo Hospital.

Consultation with Dennis Barber about elevated prostate-specific antigen (PSA)

Dr. Obioma Kalu sat in his clinic room at Quintomo Hospital, ready to begin his morning session. The room was bright and welcoming, with the gentle hum of medical equipment in the background. As the first patient, Dennis Barber, entered the room, Dr. Kalu greeted him warmly.

"Good morning, Mr. Barber. Please have a seat," Dr. Kalu said, motioning towards the patient's seat. "It's a pleasure to meet you."

Dennis smiled and settled into the chair. "Likewise, Dr. Kalu. I must say, it's quite impressive to see an African surgeon at the top of his field here in Quintomo Hospital."

Dr. Kalu chuckled, appreciating Dennis's curiosity and enthusiasm. "Thank you, Mr. Barber. I'm honoured to be part of the medical team here. Please, call me Obi."

Dennis nodded, his eyes still filled with curiosity. "Obi it is then. So, where are you from, Obi?"

Obi replied with a warm smile, "I'm originally from Nigeria. But I've been fortunate to have trained and worked in various parts of the world, including the United States."

Dennis leaned forward, his curiosity piqued. "Are you a refugee from Africa, then?"

Obi shook his head, maintaining his calm demeanour. "No, I'm not a refugee. I was born in Nigeria and pursued my medical education and training there. Later, I had the opportunity to further my career in the United States before joining the esteemed team at Quintomo Hospital."

Dennis nodded, seemingly satisfied with Obi's response. "That's quite impressive. Australia is lucky to have you. Speaking of which, what do you think of our country?"

Obi's face lit up with genuine appreciation. "Australia is a beautiful country, and I've been enjoying my time here. The people are incredibly friendly and laid-back. I feel grateful to be a part of Quintomo Hospital and to work alongside such talented and dedicated colleagues."

Dennis couldn't help but agree. "Absolutely! Australians are known for their hospitality. It's great to have professionals like you contributing to our healthcare system. Nigerians are quite remarkable too, I must say."

Dr Kalu nodded, his eyes reflecting pride. "Thank you, Dennis. Nigerians are a resilient and vibrant people, and I'm proud of my heritage. We have a rich culture and a history of accomplishments in various fields, including medicine."

As the conversation transitioned towards the purpose of the consultation, Jason Barber, Dennis' son, interjected gently. "Dad, shall we proceed with the consultation?"

Dennis nodded, his earlier curiosity now directed towards his own health concerns. Dr. Kalu listened attentively, his compassionate demeanour putting Dennis at ease.

"I've been having some concerns about my prostate health, and my son, Jason, who is an AI Engineer at this hospital, suggested I seek a second opinion from you."

Dr. Kalu nodded, empathizing with Dennis's worries. "I'm here to help, Mr. Barber. Let's discuss your concerns and explore the best course of action together."

Dennis's eyes flickered with uncertainty as he hesitated for a moment. "One thing I'm unsure about is how to share my medical information from my doctors back in Canberra. Will they have access to my records here?"

Dr. Kalu reassured him, leaning forward with a comforting smile. "Don't worry, Mr. Barber. Most hospitals in major cities participate in a government mandated nationwide network that allows for the secure sharing of patient information. In fact, with the help of Primeo, our AI system, we have already pulled your data and summarized it nicely for our review."

Relief washed over Dennis's face as he realized the convenience and efficiency of the system. "That's reassuring to hear, Obi. I'm glad that my medical history will be readily available for your evaluation."

Dr. Kalu nodded and began explaining the details of Dennis's previous consultations. "Based on the information we have, your urologist in Canberra is Dr. Elizabeth Sullivan, and your primary care physician is Dr. Mark Reynolds. During your recent consultation a month ago, it was decided to continue monitoring your PSA levels despite the slight increase over the past six months."

Dennis's son, Jason, interjected with a hint of concern. "But Dad's worried about the recent increase in his PSA level. He's concerned it could be an indication of prostate cancer."

Dr. Kalu nodded, understanding Dennis's apprehension. "I understand your worries. However, it's important to consider various factors, including your medical history, family history, and the rate of increase in your PSA level." "Let us check in with our prostate cancer risk prediction algorithm." Dr Kalu added. Dennis and Jason waited for what seemed like forever for the algorithm's recommendation. Dr Kalu continued, "Our prediction algorithm, which analysed your medical history along with genetic, molecular and imaging data, indicates a low risk of prostate cancer at this time. Continued observation and monitoring are recommended."

Dennis let out a sigh of relief, the weight of uncertainty gradually lifting from his shoulders. "That's good news to hear, Obi. I was quite anxious about the possibility of cancer."

Dr. Kalu smiled, his reassuring presence calming the room. "I'm glad I could provide you with some reassurance, Mr. Barber. Prostate health is an important matter, and we'll continue to monitor your condition closely. Should any changes or concerns arise, please don't hesitate to reach out. Our goal is to ensure your well-being and peace of mind."

As the consultation came to an end, Dennis and his son exchanged grateful smiles. They stood up from their chairs, thanking Dr. Kalu for his expertise and compassionate care.

"Thank you, Obi," Dennis said, extending his hand for a handshake. "You've eased our worries and given us a positive outlook. We appreciate your thoroughness and expertise."

Dr. Kalu shook Dennis's hand warmly. "You're most welcome, Mr. Barber. Cheers!" Dennis and Jason left the clinic room with a sense of relief and gratitude. They felt confident in Dr. Kalu's expertise and the comprehensive approach taken to address Dennis's prostate health concerns. As they walked out into the bustling hospital corridors, they couldn't help but feel optimistic about the future, knowing they had a caring and knowledgeable urologist on their side.

Ralph Gordon recovers from robot-assisted prostate surgery

Dr. Kalu entered the consultation room, ready to see his next patient, Ralph Gordon, who was recovering from a robot-assisted total prostatectomy. Ralph, a 71-year-old man, greeted Dr. Kalu with a warm smile.

"Good morning, Mr. Gordon. How are you feeling today?" Dr. Kalu asked, taking a seat across from Ralph.

Ralph grinned. "I'm feeling great, Dr. Kalu. It's amazing how far I've come since the diagnosis. I still can't believe it all started with an insect bite and ended up with prostate cancer."

Dr. Kalu nodded, understanding the gravity of Ralph's journey. "Indeed, it's quite remarkable how things unfolded. But I'm glad we were able to diagnose your prostate cancer early and swiftly proceed with the necessary treatment."

Ralph chuckled softly. "Lucky me, right? Who would have thought that a little bug bite could lead to such a journey of cellulitis then a blood clot in the lung leading to cardiac arrest then ultimately uncovering the prostate cancer? But I'm grateful for the excellent care I received from the emergency department to the intensive care unit and, of course, from you during my surgery."

Dr. Kalu smiled warmly, acknowledging Ralph's appreciation. "Thank you, Mr. Gordon. It was a team effort, and I'm glad to see that you're recovering well after the surgery. How have you been feeling overall?"

Ralph leaned back in his chair, looking visibly relieved. "Honestly, Dr. Kalu, I've been feeling great. I haven't experienced any complications or issues since the surgery. The recovery process has been smoother than I expected."

Dr. Kalu nodded, pleased with Ralph's progress. "That's wonderful to hear, Mr. Gordon. Your recovery is progressing as expected. We conducted regular follow-up tests, and the results show that your post-operative healing is on track. Your body has responded well to the treatment."

Ralph sighed with relief, a genuine smile spreading across his face. "I can't express how grateful I am, Dr. Kalu. You and your team have given me a new lease on life. I feel optimistic about the future."

Dr. Kalu reached out, placing a reassuring hand on Ralph's arm. "That's the goal, Mr. Gordon. Our primary aim is to provide you with the best care and improve your quality of life. Remember, I'll be here every step of the way to support you throughout your recovery journey."

Ralph nodded, a sense of gratitude evident in his eyes. "Thank you, Dr. Kalu. I truly appreciate all that you've done for me. I'm looking forward to moving forward with renewed health and vitality."

Dr. Kalu smiled, his dedication to his patients shining through. "You're most welcome, Mr. Gordon. We'll continue to monitor your progress and provide any necessary support. If you have any questions or concerns at any point, please don't hesitate to reach out. Your well-being is our priority."

As Ralph stood up, ready to leave the consultation room, he couldn't help but reflect on his journey. "Who would have thought a bug bite would lead me to such a remarkable doctor and a life-saving surgery? I feel truly blessed."

Dr. Kalu's eyes twinkled with warmth as he bid farewell to Ralph. "It's been a privilege to be part of your journey, Mr. Gordon. Take care, and remember, we're here for you every step of the way."

Ralph nodded gratefully, his heart filled with hope and appreciation. "Thank you, Dr. Kalu. I'll see you at the next follow-up appointment."

With that, Ralph left the consultation room, filled with gratitude and a renewed sense of hope for the future. Dr. Kalu, too, felt a sense of fulfillment, knowing that he had made a positive impact on Ralph's life and helped him navigate through a challenging journey. As he prepared for his next patient, he carried Ralph's story with him, a reminder of the profound impact that urology and compassionate care can have on the lives of patients.

An urgent video call from a patient's son

As Dr. Kalu wrapped up his day in the office, he focused on reviewing and signing off the status reports generated by Primeo. The room was filled with the gentle hum of medical equipment, creating a calm ambiance. He diligently attended to his emails and phone calls, ensuring all necessary tasks were addressed.

In the midst of his routine, Primeo interrupted with an incoming urgent video call from Tom, the next of kin of one of

Dr. Kalu's patients. Dr. Kalu granted permission to connect the call.

Tom's somber face appeared on the screen, and a sense of unease washed over Dr. Kalu. Without any preamble, Tom's voice quivered as he delivered the heart-wrenching news. "Obi, I'm sorry to inform you... my father has passed away."

CHAPTER 24

TIME OF DEATH PREDICTION ALGORITHM

Austin Brewer, an 85-year-old man, had faced the challenging battle against bladder cancer that had spread to his lymph nodes, liver, and bones. His journey began five years ago when he was diagnosed with Stage IV bladder cancer. Despite the advanced stage, Austin underwent a series of appropriate treatments to control the disease.

His treatment regimen included a combination of chemotherapy, radiation therapy, surgery to remove the primary tumour, and immunotherapy. Over time, Austin showed remarkable resilience and managed to recover with a decent quality of life. He continued to receive ongoing treatment to prevent the progression of his cancer, attending regular check-ups and monitoring sessions with his medical team.

In addition to his bladder cancer, Austin had a complex medical history. He had been living with atrial fibrillation, a heart rhythm disorder, for which he required oral anticoagulation. He also had a history of hypertension and heart failure, which required careful management alongside his cancer treatment. Throughout the years, Austin had experienced ups and downs, but the positive moments seemed to outweigh the challenges he faced.

Recently, Austin's condition took a turn for the worse, and his disease began to progress despite the previous treatments. It was at this point that he was referred to Quintomo Hospital, where Dr. Obioma Kalu, a skilled and compassionate urologist, took over his care. Dr. Kalu became an integral part of Austin's medical team, working closely with the multidisciplinary uro-oncology team to determine the best course of action.

Unfortunately, Austin's cancer had become resistant to conventional treatments, including participation in clinical trials involving gene-based immunotherapies. After thorough discussion and consideration during a multidisciplinary uro-oncology meeting, the decision was made to transition Austin's care to palliative care. This decision aimed to provide him with the best possible quality of life and symptom management as his disease progressed.

Austin's Final Journey

Austin's health had taken a sudden downturn, leading to his admission to the hospital. The once vibrant and resilient man now struggled with frank bloody urine, a distressing symptom of his advanced bladder cancer. Dr. Obioma Kalu, aware of the urgency of the situation, promptly performed a cystoscopy to assess the extent of the tumour progression. The findings were disheartening— the cancer had spread further within the bladder.

The severity of the bleeding necessitated ongoing irrigation, and as a precautionary measure, Austin's anticoagulation medication was discontinued upon admission. The medical team understood the delicate balance between managing his bleeding and ensuring his safety.

During this challenging time, Austin and his family made a decision that would guide their journey. They chose to opt into the time of death prediction algorithm, which could forecast the approximate date and time of his passing, allowing them to prepare and ensure his wishes were met. It was Austin's desire to be surrounded by his loved ones in the hospital during his final moments.

To provide Austin with the best possible care and support, he was placed on an end-of-life pathway. The palliative care team worked in harmony with the urology team, nursing staff, and other allied health professionals to ensure his comfort and dignity were upheld. The hospital became a haven for Austin and his family, a place where they could find solace amidst the emotional turbulence.

Despite the challenging circumstances, Austin's sons from Canada and the United Kingdom were able to travel to be by his side in his final days. It was a testament to the strength of their bond and the significance of Austin's presence in their lives.

In the tender hours leading up to his passing, Austin peacefully departed, surrounded by his family as he had wished. The algorithm's prediction had indicated a specific time, but Austin's journey came to a close two hours prior to the projected moment. His family found solace in knowing that he was at peace and that they were able to share those precious final moments together.

In the aftermath of Austin's passing, his family expressed their gratitude for the exceptional care provided by the urology team, the palliative care team, and the dedicated nurses and allied health professionals who had supported them throughout their journey. Their appreciation resonated deeply within the hospital walls, reaffirming the importance of compassion, empathy, and teamwork in delivering end-of-life care.

Following Austin's passing, his son, Tom, rang his attending urologist, Dr. Obioma Kalu to express his gratitude for the care his dad received. Without any preamble, Tom's voice quivered as he delivered the heart-wrenching news. "Obi, I'm sorry to inform you... my father has passed away."

"I wanted to personally thank you for everything you did for my father during his final days," Tom said, his voice filled with genuine appreciation.

Dr. Kalu smiled warmly, grateful for the opportunity to connect with Austin's family. "You're very welcome, Tom. I'm glad I could be there for your father and provide the care he deserved."

Tom took a deep breath before continuing, "I also wanted to mention how much the time of death prediction algorithm meant to our family. It helped us prepare ourselves emotionally and be there for Dad during his final hours."

Dr. Kalu nodded, understanding the significance of the algorithm's role in their journey. "I'm glad it provided some comfort and helped you plan as a family. It's a tool we use to assist patients and their families in making important decisions and ensuring their wishes are honoured."

Tom's voice grew softer as he spoke, "It truly did make a difference, Obi. Knowing the approximate time allowed us to gather our loved ones, travel from afar, and be by Dad's side during his final moments. It brought us all a sense of peace."

Dr. Kalu listened attentively, his heart touched by the impact their experience had on the family. "I'm honoured to have been part of that journey with you. It's moments like these that remind me why I chose this path—to provide compassionate care and support during difficult times."

Tom expressed his gratitude once more before ending the call, "Thank you, Obi. Your kindness and expertise meant the world to us. We will forever be grateful for the care you provided."

Dr. Kalu's voice was filled with sincerity as he replied, "It was my privilege, Tom. Your father was a remarkable man, and it was an honour to be there for him and your family. If there's ever anything else I can do, please don't hesitate to reach out."

As they bid each other farewell, Dr. Kalu couldn't help but reflect on the impact he had made in their lives. The time of death prediction algorithm had not only guided their journey but had also instilled a sense of comfort and control during a time of uncertainty. It served as a reminder of the power of technology in enhancing patient care and supporting families in their time of need.

In the days that followed, Dr. Kalu carried with him the gratitude expressed by Tom and his family. Their experience reinforced his commitment to providing compassionate care, incorporating innovative tools, and ensuring that patients and their families receive the support they deserve during life's most challenging moments.

Jesse Dougherty is admitted with severe sepsis

Jesse Dougherty, an 82-year-old retired plumber, was accompanied by his wife, Peggy, and his son, Luke, as he was admitted to the emergency department. Jesse's health had taken a downturn as he presented with symptoms of community-acquired pneumonia. His medical history included several significant conditions, such as ischaemic heart disease, a previous history of coronary bypass surgery, treated bowel cancer, hypertension, type 2 diabetes mellitus, hypercholesterolemia, peripheral vascular disease, and previous strokes. His mobility was limited, relying on a wheelchair, and he required assistance with most activities of daily living.

As Jesse was being attended to by Dr. Moses Levi, an advanced trainee registrar in the department of general medicine, took charge of his care. Moses reviewed Jesse's medical records and assessed his condition. The findings were concerning, indicating that Jesse had severe sepsis resulting from community-acquired pneumonia. The infection had taken a toll on his body, affecting his kidneys and causing his brain to become delirious, resulting in confused conversations.

Moses approached Jesse, offering reassurance and compassion. "Mr. Dougherty, I'm Dr. Moses Levi, and I will be overseeing your care during your time in the hospital. Your current symptoms and test results indicate that you have severe sepsis from pneumonia. This infection has affected your kidneys and caused some confusion in your thinking."

Jesse blinked, trying to process the information as his wife and son listened intently. Moses continued, explaining the plan of care in simpler terms. "We will be providing you with antibiotics to fight the infection, intravenous fluids to support your kidneys, and other medications to stabilize your condition. We will also closely monitor your vital signs and work towards managing the infection and improving your overall well-being."

Peggy leaned forward, her concern evident. "Will he be alright, doctor?"

Moses met her gaze with empathy. "We are doing everything we can to treat the infection and support his recovery. Jesse is in good hands, and we will monitor his progress closely. It's important to remember that the road to recovery may take some time, and we will be here every step of the way."

Luke, Jesse's son, sought clarification. "What can we do to help him?"

Moses nodded appreciatively. "Your presence and support are invaluable. Being by his side, providing emotional comfort, and helping him with daily activities will make a significant difference. We will also keep you updated on any changes in his condition and involve you in the decision-making process regarding his care."

Moses took a moment to collect his thoughts before continuing the conversation with Jesse's family. He understood the sensitivity of the topic and wanted to provide them with clear and compassionate explanations. He turned towards Peggy and Luke, addressing them with a gentle yet assured tone.

"Peggy, Luke, I wanted to discuss something important with you regarding Jesse's care. Our hospital offers an opt-in program called the Time of Death Prediction Algorithm, which provides an estimation of when a patient may pass away. This program is designed for patients who are very unwell or at high risk of dying during their admission."

Peggy's eyes widened with curiosity and concern. "What does that mean, doctor?"

Moses nodded, acknowledging her reaction. "I understand this may sound overwhelming, but it's important to consider Jesse's wishes and his current condition. I have reviewed his advanced directive, and it states that he would prefer not to receive resuscitation and instead be kept comfortable if he is not responding to treatment. As the enduring power of attorney, your role in decision-making is vital."

Peggy nodded, her voice filled with a mix of determination and concern. "Yes, we have discussed this as a family and want to honour Jesse's wishes. We want him to be comfortable."

Moses reassured her, offering support and understanding. "Thank you for clarifying that, Peggy. I want to assure you that we will be in close communication with you, keeping you informed every step of the way. Our priority is to honour Jesse's wishes and provide him with the best possible care."

Luke leaned forward, his brows furrowed with curiosity. "But how does this Time of Death Prediction Algorithm work? Is it accurate?"

Moses nodded, appreciating Luke's thoughtful question. "The algorithm takes into account multiple data points, including Jesse's medical history, investigation results, genetics, and information from sensors. It provides an approximate estimation of the time Jesse may pass away. During active treatment, it can indicate how unwell a patient is and their current trajectory. As the patient improves, the prediction can shift to indicate a lower risk of dying during the admission, which is a positive outcome."

Peggy interjected, her voice tinged with concern. "What if the algorithm predicts a shorter time to death? What does that mean for Jesse?"

Moses's empathetic gaze met Peggy's. "If the prediction suggests a shorter time to death and Jesse is not responding to appropriate treatment, it can help guide us in adjusting our approach. It might indicate that active treatment may not be achieving the desired results. In such cases, the algorithm's information can assist both the family and the medical team in deciding to focus on comfort measures and providing Jesse with the best possible care."

Luke, with a mix of curiosity and pragmatism, asked, "How does this information help us make decisions?"

Moses paused for a moment, collecting his thoughts before responding. "The Time of Death Prediction Algorithm works in conjunction with our AI assistant, Primeo. It considers the patient's goals of care and clinical trajectory. Based on its analysis, it can make recommendations for the best course of action. The information provided can support the family and medical team in making informed decisions that align with Jesse's wishes and ensure his comfort and well-being."

After a moment of contemplation, Peggy turned to Moses with determination. "We want to opt-in to the program. Having this information will help us plan and ensure Jesse receives the care he desires."

Moses nodded, acknowledging their decision. "I understand, Peggy. Your choice is respected, and I commend your commitment to honouring Jesse's wishes. We will proceed accordingly, and Primeo will provide us with the necessary information and recommendations throughout Jesse's care."

With the decision made, Moses felt a sense of relief, knowing that the family was actively engaged in the decision-making process. With their consent to opt-in to the Time of Death Prediction Algorithm, he could provide them with the support and guidance they needed in navigating Jesse's care.

Jesse Dougherty deteriorates

Moses, in the midst of attending to other patients in the busy emergency department, felt his Primeo Glass vibrate with a notification. As he glanced at the message, his heart sank. It was an emergency call for Jesse Dougherty, the patient he had admitted earlier.

Without wasting a moment, Moses rushed to a quiet room in the emergency department. He connected with the emergency team and watched the live video feed from the scene via his Primeo Glass. The atmosphere was tense, and he could see the worry etched on the faces of Jesse's family members who stood by his bedside.

Chapter 25

THE ART OF MEDICINE: REFLECTIONS ON PROGNOSIS AND CARE

The team of doctors and nurses worked swiftly to assess Jesse, their movements choreographed with precision. Moses joined in, utilizing the information provided by Primeo to quickly grasp the situation. Jesse's deteriorating condition demanded immediate intervention.

Moses leaned in closer, his voice calm but determined. "Let's administer high-flow oxygen," he directed, while simultaneously observing the readings on the monitors and communicating with the team via smart speakers with his video displaying on a monitor.

As the team diligently carried out the necessary interventions, Jesse's family members held onto each other, their eyes filled with concern. Moses, via live video, shared updates with them, providing reassurance amidst the chaos. "We're doing everything we can to stabilize Jesse and improve his condition. Please stay strong and hang in there."

Minutes felt like hours as the team worked tirelessly to stabilize Jesse's deteriorating vital signs. Finally, a collective sigh of relief echoed through the room as Jesse's condition began to stabilize. The emergency team's efforts had paid off, at least for the moment.

Moses took a step back, his gaze shifting from the monitors to Jesse's family. With genuine empathy in his voice, he said, "He's stable for now, but we'll continue to closely monitor him. You've all been an incredible support for him throughout this journey, and your presence is making a significant difference. I will come to the ward and chat with you shortly." Jesse's family thanked Moses and the emergency team for their efforts.

Moses made his way to Jesse's room on the medical ward, his mind still reeling from the intensity of the emergency situation. As he entered the room, he could feel the weight of anticipation and worry that hung in the air. The room was dimly lit, the soft glow of the bedside lamp casting a warm and gentle light.

Jesse's family was gathered around his bed, their faces etched with concern and exhaustion. The beeping of medical equipment provided a constant backdrop to the room, a reminder of the fragile nature of Jesse's condition. The atmosphere was tense, yet Moses could sense a glimmer of hope in the room, a shared resilience among the family members.

Moses approached the bed, his presence calming the room ever so slightly. He spoke softly, his voice laced with both compassion and determination.

Moses took a deep breath, his gaze shifting from the medical monitors to Jesse's family gathered by his bedside. He knew that open and honest communication was crucial in moments like these.

"Peggy, Luke, I want to update you on Jesse's condition," Moses began, his voice filled with compassion. "His condition has worsened, and the time to death prediction algorithm now indicates a timeframe of approximately three days. However, it's important to remember that this is just an estimate, and every patient is unique."

Peggy clutched her husband's hand, her eyes welling up with tears. "Is there anything more you can do, Dr. Levi? We don't want him to suffer."

Moses nodded, understanding their concerns. "We've taken simple measures to address his immediate needs, keeping in line with Jesse's wishes and advanced directive. Our focus now is to closely monitor his response to treatment and ensure he remains as comfortable as possible."

Luke spoke up, his voice filled with a mix of sadness and determination. "We trust you, Dr. Levi. We know you'll do what's best for Dad. Please keep us informed about any changes."

Moses placed a comforting hand on Luke's shoulder. "Absolutely, Luke. We're committed to providing your father with the highest level of care. We'll continue to monitor his condition closely and keep you updated on any developments or changes in his treatment plan."

Peggy wiped away a tear, her voice filled with both gratitude and concern. "If he doesn't respond to treatment, what then? We don't want to prolong his suffering."

Moses maintained his gentle composure, offering reassurance. "I understand your concerns, Peggy. We respect Jesse's wishes, and if his condition doesn't improve or if further treatment becomes futile, we will shift the goals of care towards comfort and focus on ensuring he's as pain-free as possible."

Peggy nodded, a mix of relief and sadness crossing her face. "Thank you, Dr. Levi. We appreciate your honesty and compassion. All we want is for Jesse to be at peace."

Moses nodded in understanding. "That's our goal too, Peggy. We're here to support both Jesse and your family through this difficult time. Please remember that you can reach out to us at any moment, and we'll be by your side every step of the way."

Jesse's wife, Peggy, reached out and held Moses' hand tightly. Her voice trembled with gratitude as she said, "Thank you, Dr. Levi. We appreciate everything you and the team are doing for Jesse."

As the conversation came to a close, Moses lingered for a moment longer, offering a silent moment of solace for the family. He knew that their journey was far from over, but he remained steadfast in his commitment to providing compassionate care.

With the emergency situation under control, Moses returned to his duties in the bustling emergency department. As he attended to other patients, Moses couldn't help but reflect on the unpredictable nature of medicine and the importance of relying on the collective efforts of both human expertise and technological advancements. With renewed determination, he continued his work, ready to face whatever challenges the night may bring.

General Medicine ward round with Dr Kirra Nakata

Dr. Kirra Nakata, a seasoned Acute and General Physician, led the ward round the next morning with her team of resident doctors, medical students, and an attentive ward nurse. Among the patients they reviewed was Mr. Harold Thompson, a 91-year-old gentleman who had been admitted following a fall caused by postural hypotension due to his medications and dehydration.

Mr. Thompson was a sprightly man, with a gentle smile and a twinkle in his eye. He had a remarkable medical history, having been diagnosed with hypertension, type 2 diabetes mellitus, and chronic kidney disease. He was on a cocktail of medications, including a calcium channel blocker for his blood pressure, metformin for diabetes, and a diuretic for fluid management. His medical problems were complex, but his spirit remained resilient.

As the team gathered around the bedside, Primeo, the hospital's AI assistant, provided a concise summary of Mr. Thompson's case, including his recent fall, the precipitating factors, and his current management plan. Dr. Moses Levi, eager to share more context, chimed in with additional anecdotes about Mr. Thompson's medication adherence and his independent lifestyle prior to the fall.

Dr. Nakata took a moment to review Mr. Thompson's medical records, considering the delicate balance between medication optimization and the potential adverse effects. She then proceeded to engage Mr. Thompson in a conversation, taking a brief history and performing a thorough physical examination.

"Mr. Thompson, I understand you had a bit of a fall yesterday," Dr. Nakata began, her tone warm and reassuring. "We've been looking into the reasons behind it, and it seems like your blood pressure dropped when you stood up quickly, causing you to lose your balance. We call this postural hypotension."

Mr. Thompson nodded, his face reflecting a mix of understanding and concern. "Yes, I remember feeling lightheaded and dizzy before I fell. It's been a bit of a rough patch lately."

Dr. Nakata placed a reassuring hand on his arm. "I can imagine, Mr. Thompson. We're working on getting you back on your feet. We've made some adjustments to your medications, including stopping one of your blood pressure tablets temporarily. Additionally, we've started you on intravenous fluids to rehydrate you."

Mr. Thompson smiled gratefully. "Thank you, Doctor. I trust you know what's best. It's a relief to have such caring doctors looking after me."

Dr. Nakata returned the smile, her empathy shining through her words. "We're here to support you every step of the way, Mr. Thompson. We want to ensure that your blood pressure remains stable and that you regain your strength. With the improvement in your postural hypotension, our plan is to discharge you to the virtual ward for another day or two. This way, we can keep a close eye on you while you continue to recover at home."

Cindy, the ward nurse, who had been listening attentively, chimed in. "We'll make sure you have all the support you need at home, Mr. Thompson. Our virtual ward team will be in regular contact, monitoring your progress and assisting with any concerns you may have."

Mr. Thompson expressed his gratitude to the entire team, his eyes welling up with emotion. "I feel fortunate to be in such capable hands. Thank you all for your dedication and care."

Dr. Nakata nodded, a sense of satisfaction spreading through her. "You're most welcome, Mr. Thompson. It's patients like you who inspire us to do what we do. We'll continue working together to ensure your recovery is smooth and successful."

With that, the team moved on to the next patient, their focus unwavering as they embarked on their rounds, providing compassionate care and expertise to each individual under their care.

Jesse Dougherty receives specialist review

The medical team entered Jesse Dougherty's room, ready to discuss his condition and plan for his ongoing care. Dr. Kirra Nakata, the attending physician, led the discussion, with Dr. Moses Levi, other resident doctors, medical students, and a ward nurse in attendance. Jesse's case was complex, with a history of severe sepsis secondary to community-acquired pneumonia and multiple comorbidities.

As the team gathered around Jesse's bed, Primeo, the AI assistant, began presenting a summary of the night's events. "Jesse Dougherty is an 82 year old man admitted with severe sepsis secondary to community acquired pneumonia. Last night, Jesse experienced a significant deterioration, presenting with hypotension and hypoxia," Primeo reported. "The emergency team was called in and successfully stabilized him, administering intravenous fluid boluses and high flow nasal oxygen to support his vital signs."

Dr. Nakata nodded, acknowledging the gravity of Jesse's condition. She turned her attention to Primeo. "Thank you for the update. Please continue."

Primeo continued, "Despite the stabilization, the time of death prediction algorithm indicated a predicted time of death within 3 days. However, it's important to remember that this prediction has a margin of error and is subject to change based on the patient's response to treatment."

Dr. Nakata looked at the team, considering the implications of the prediction. "The predicted time of death is a valuable piece of information, but it should not solely dictate our approach to care. Our goal is to provide the best possible care and support for Jesse and his family."

Dr. Moses Levi, who had been instrumental in Jesse's initial stabilization, chimed in. "Agreed, Dr. Nakata. We need to provide comprehensive and compassionate care, taking into account Jesse's wishes and his family's input."

Dr. Nakata turned her attention to Jesse's wife, Peggy, and his son, Luke, who were present in the room. "Mrs. Dougherty, Luke, I understand this may be a difficult time for you. I want to assure you that our priority is Jesse's well-being. We will closely monitor his condition and keep you informed every step of the way."

After thoroughly assessing Jesse's condition and carefully reviewing his medical records, Dr. Nakata delivered the much-awaited news to the patient's family. With a smile on her face, she announced, "I'm pleased to inform you that Jesse's condition is showing signs of improvement. The time of death prediction algorithm has been revised, and it now indicates a predicted time of death within 7 days."

The atmosphere in the room instantly shifted, and a wave of relief washed over the faces of Jesse's wife, Peggy, and his son, Luke. Their worries began to subside, replaced by a glimmer of hope. Luke couldn't help but let out a sigh of relief and exclaimed,

"That's wonderful news, Dr. Nakata! We were so worried, but this gives us some much-needed hope."

Dr. Nakata nodded, sharing in their joy. "Yes, it's indeed a positive development. Jesse's response to the treatment has been encouraging, and we will continue to closely monitor his progress. While we cannot predict the future with absolute certainty, we are committed to providing the best possible care and support for Jesse."

Peggy, Jesse's devoted wife, clasped her hands together, her eyes welling up with tears of gratitude. "Thank you, Dr. Nakata, and the entire medical team. We appreciate your dedication and the care you have shown to Jesse."

Dr. Nakata reassured them, her compassionate tone filled with empathy. "You're most welcome, Peggy. We understand the challenges you and your family have been facing. Our team is here for you every step of the way, providing support, and ensuring that Jesse's comfort and well-being are prioritized."

With the revised prediction offering a glimmer of hope, the room became filled with a renewed sense of optimism. The family members embraced each other, their joy radiating through the room. Dr. Nakata and the medical team left the room, knowing that their continued efforts would help Jesse on his path to recovery.

In the days that followed, the family remained hopeful, cherishing each moment with Jesse. The revised time of death prediction served as a beacon of light, a reminder that there was still time to cherish and make memories together. The medical team diligently carried out their duties, providing attentive care, and ensuring that Jesse's comfort and well-being were maintained.

The journey was far from over, but with every passing day, the family's hope grew stronger. The revised prediction became a testament to the power of medical intervention, support, and the resilience of the human spirit. Together, they faced the future with newfound strength and optimism, embracing each day as a precious gift.

Dr. Kirra Nakata reflects on the time of death prediction algorithm

Dr. Nakata settled into her office chair, the events of the ward round still fresh in her mind. She took a moment to reflect on the day's encounters and the information provided by the time of death prediction algorithm. As she reviewed the patient's cases and planned for the upcoming clinic, a mix of thoughts and emotions filled her thoughts.

The revised time of death prediction algorithm had brought a sense of relief to Jesse's family, and she understood the significance of this information. However, as an experienced physician, Dr. Nakata couldn't help but ponder the true value of such predictions. While the algorithm provided a specific timeframe, she had always approached end-of-life discussions with patients and their families using a more generalized approach, acknowledging the inherent uncertainty in predicting the future.

In her experience, she had often witnessed doctors offering a wider range of potential outcomes when discussing prognosis with patients nearing the end of life. It allowed for the variability and unpredictability of each individual's journey. Dr. Nakata believed in the importance of open and honest communication, providing support and comfort while navigating the complexities of illness.

Although the time of death prediction algorithm could be a valuable tool, she couldn't help but maintain a somewhat neutral stance towards it. She understood its potential benefits, particularly for patients and families who sought specific information and appreciated a more precise prediction. However, she recognized that her clinical judgment and expertise, coupled with compassionate care, were fundamental in guiding patients through their medical journeys.

As she organized her thoughts for the clinic ahead, Dr. Nakata reminded herself to approach each patient with empathy, understanding, and a commitment to providing the best care possible. While the algorithm was another tool in her medical arsenal, it would never replace the human connection, the art of medicine, and the intuitive judgment she had honed throughout her career.

With a determined mindset, Dr. Nakata continued her preparations, ready to face the challenges and uncertainties that lay ahead. She recognized the importance of blending the advances of technology with her clinical expertise, knowing that the ultimate goal was to provide comprehensive, compassionate care to her patients.

As she left her office, she carried with her the belief that medicine was a delicate balance between science and humanity. The time of death prediction algorithm was a piece of the puzzle, offering insights and guidance, but it would always be her duty as a physician to navigate the complexities of illness with care, empathy, and a deep understanding of each patient's unique journey.

Chapter 26

A Tense Breakthrough: Proposals, Reluctant Agreements, and Lingering Doubts

The conference room at Quintomo Hospital buzzed with activity as the legal team gathered to prepare for the upcoming mediation meeting regarding Joshua Lee's case. Andy Sumner, the Director of Legal Services, led the team, joined by his colleagues, Mark Johnson and Laura Bennett.

Andy cleared his throat, signalling the start of the meeting. "Good morning, everyone. Let's begin by reviewing the case of Dr. Joshua Lee. As you all know, he was stood down after his sensors detected a high blood alcohol level. Our goal is to present a strong defence during the mediation and protect the reputation of Quintomo Hospital."

Mark, a seasoned lawyer with a no-nonsense demeanour, chimed in, "Absolutely, Andy. We have gathered substantial evidence to support our position. First and foremost, we have the result of the sensors, which clearly indicated an elevated blood alcohol level. We can also present the quality testing and calibration report of the sensors to establish their reliability."

Laura, a meticulous and detail-oriented lawyer, added, "Furthermore, we have evidence of Joshua's attempt to tamper with the sensors on the morning of the incident. This suggests an awareness of his actions and an attempt to manipulate the results."

Andy nodded, acknowledging their points. "Indeed, these are crucial pieces of evidence. Additionally, we have gathered information that indicates Joshua had been out drinking the previous evening. While he claims he didn't drink excessively, we need to question whether he had more to drink at home before bed. It will be important to establish the timeline and determine the accuracy of his statement."

Mark leaned forward, his voice filled with determination. "We should also emphasize the potential consequences of Joshua's actions. As a medical professional, he is expected to uphold high standards and prioritize patient safety. Any impairment, whether due to alcohol or other factors, could compromise his ability to provide optimal care."

Laura interjected, her voice measured and confident. "Let's not forget to address the issue of trust. The hospital relies on its staff to maintain a culture of professionalism and integrity. If patients and colleagues can't trust Joshua to make sound decisions, it raises concerns about his fitness to practice."

Andy nodded in agreement. "Excellent points, Mark and Laura. We need to carefully construct our arguments and counter any claims Joshua's lawyers may present during mediation. Our goal is to demonstrate that the hospital's actions were reasonable, given the evidence at hand, and that Quintomo's commitment to patient safety and professionalism guided our decision-making process."

The legal team continued their intense discussion, analysing every aspect of the case and brainstorming strategies for the upcoming mediation meeting. They meticulously prepared their arguments, anticipating potential objections and challenges from Joshua's legal team.

In the conference room, the atmosphere was charged with focused determination as Andy, Mark, and Laura worked together to ensure Quintomo Hospital's position was well-represented. Their extensive preparation and commitment to delivering a strong defence set the stage for a critical and impactful mediation session.

Joshua Lee's mediation terms

As the mediation meeting commenced, Joshua sat alongside his legal team, facing the representatives of Quintomo Hospital. The atmosphere was tense, with both sides prepared to present their arguments and negotiate a resolution. Joshua's legal team took the floor, led by Michael Pearson, to present their terms of mediation.

Michael began by asserting Joshua's innocence, emphasizing that the sensors at Quintomo Hospital had made a grave mistake in detecting a high blood alcohol level. He highlighted the emotional and financial toll the ordeal had taken on Joshua, as well as the significant damage to his professional reputation, including the disciplinary measures imposed by AHPRA.

"With all due respect," Michael stated firmly, "we firmly believe that Joshua is innocent of any wrongdoing. The hospital's sensors have clearly malfunctioned, leading to a false accusation against our client. This has caused him immense emotional distress, financial burden, and has severely impacted his professional standing within the medical community."

Sarah Robertson, another member of Joshua's legal team, interjected, "Given the gravity of the situation, we propose a fair and reasonable resolution. Quintomo Hospital should issue a public apology, acknowledging the error of their sensors and the subsequent damage caused to Joshua's reputation. This public acknowledgment is crucial to reversing the disciplinary measures imposed by AHPRA."

Joshua's voice resonated with determination as he continued, "In return for the hospital's apology and acknowledgment, we are prepared to drop our planned lawsuit for unlawful tracking, monitoring, and the damage inflicted upon my reputation. We believe that this outcome will provide some measure of justice and allow us all to move forward."

Across the table, the representatives of Quintomo Hospital listened intently, their expressions reflecting a mixture of contemplation and evaluation. The terms presented by Joshua's legal team held significant weight, acknowledging the impact on his personal and professional life, while providing an opportunity for resolution without prolonged legal battles.

The room fell silent for a moment as both sides contemplated the proposed terms. It was evident that the outcome of this mediation meeting would have far-reaching implications for

Joshua, the hospital, and the individuals involved. The path to resolution was uncertain, but the shared goal of finding a fair and just outcome resonated on both sides of the table.

Andy Sumner, the leader of the hospital's legal team, leaned forward, his expression composed yet determined. He prepared to address Joshua's demands, making it clear that the hospital found none of them acceptable. He began by presenting their case.

"Ladies and gentlemen, while we appreciate the concerns raised by Dr. Joshua Lee and his legal team, we must firmly assert that the demands put forth are not acceptable to Quintomo Hospital," Andy stated, his voice steady and authoritative.

He continued, "First and foremost, it is important to clarify that Quintomo Hospital stands by the accuracy and reliability of its sensors. Our records indicate that Joshua's blood alcohol level was indeed elevated on the day in question. We have the necessary documentation, including quality testing and calibration reports, to support this claim."

Andy proceeded to outline the evidence against Joshua, highlighting not only the result of the sensors but also Joshua's attempted tampering with them that morning. He emphasized that the hospital's commitment to patient safety and the well-being of its staff necessitated strict adherence to monitoring protocols.

"While we sympathize with the stress that Joshua has endured, we must prioritize the integrity of our monitoring systems," Andy explained. "We cannot issue a public apology and acknowledge the error of our sensors when we firmly believe they functioned correctly. Doing so would undermine the trust in our technology and compromise patient safety."

He concluded by affirming the hospital's commitment to addressing the concerns raised by Joshua, but within a framework that aligns with the hospital's principles and responsibilities.

"We are open to exploring alternative avenues for resolution, ones that acknowledge the challenges Joshua has faced while also maintaining the integrity of our monitoring systems. Our primary goal is to ensure the well-being of our patients and staff, and any resolution must align with this fundamental principle," Andy stated firmly.

With his response, Andy made it clear that the hospital's legal team would not be swayed by Joshua's demands as presented. The room fell into a momentary silence, anticipation hanging in the air, as both sides grappled with the implications of the differing positions. The stage was set for further negotiations and discussions, as each team sought to find common ground and a path towards resolution.

As the mediation meeting reached a critical juncture, the legal teams of Joshua and the hospital engaged in a series of intense deliberations, assessing the feasibility and merits of proposed solutions. After careful consideration, they found common ground on certain aspects, leading to a breakthrough in their discussions.

Andy Sumner, the lead lawyer for the hospital's team, presented the proposed course of action. "In order to reach a fair resolution, we propose initiating an independent investigation to examine the accuracy and reliability of the hospital's sensor system. Additionally, we suggest exploring the possibility of arbitration to help facilitate a mutually acceptable outcome."

Joshua's initial reaction was one of scepticism and resistance. "I understand your proposal, but I firmly believe the sensors provided a false report. Regardless of what an independent investigator might say, I am convinced of my innocence."

Michael Pearson, Joshua's lawyer, empathized with his client's concerns. "Joshua, I understand your frustration and doubt. However, an independent investigation will provide an objective assessment of the sensor system. It will help establish the facts and shed light on the accuracy of the reported alcohol levels. This investigation is crucial for our case."

Reluctantly, Joshua acquiesced, realizing that an independent investigation might be his best chance to challenge the validity of the sensor system's readings. "Alright, I will agree to the proposal, but I want it to be known that I firmly believe the sensors gave a wrong report. I am willing to go through with this process to prove it."

With Joshua's reluctant agreement, the legal teams further discussed the logistics and parameters of the independent investigation. They emphasized the need for transparency, ensuring full access to data, records, and expert testimony.

"We must ensure that the investigation is thorough and impartial," Andy emphasized. "By collaborating and providing all necessary information, we can demonstrate our commitment to uncovering the truth."

Michael nodded in agreement. "Transparency is crucial. We need access to all relevant data and documents to conduct a comprehensive examination of the sensor system."

The teams also discussed the potential for arbitration, recognizing the value of an impartial third party to help facilitate a resolution. They explored the criteria for selecting an arbitrator with expertise in both medical and legal matters.

After extensive deliberation, the legal teams agreed to reconvene in two weeks' time. During this period, they would initiate the independent investigation and begin compiling their arguments for arbitration.

"We have made significant progress today," Andy stated, a note of optimism in his voice. "Let's use this momentum to move forward. We'll reconvene in two weeks to review the investigation findings and discuss potential arbitrators. Our aim is to find a fair and satisfactory resolution for all parties involved."

With a sense of determination, the legal teams concluded the meeting. Joshua, while still harbouring doubts, acknowledged the necessity of the proposed actions. He realized that by pursuing an independent investigation and exploring arbitration, he would have a chance to challenge the accuracy of the sensor system's readings and fight for his innocence.

Junior doctors worry about Joshua's future

The doctors gathered in the bustling hospital cafeteria, finding a spot to sit together at a table. The lunchtime buzz filled the air as they each unpacked their meals and settled in for a much-needed break.

Sophie took a bite of her sandwich and glanced at her colleagues. "Mmm, this chicken Caesar wrap is a lifesaver," she remarked with a smile. "What about you all? What did you bring today?"

Rachel, sipping on a cup of soup, chimed in. "I opted for a warm bowl of vegetable soup. Perfect for this chilly day."

Samantha, carefully balancing a salad bowl, nodded in agreement. "I went for a light salad with grilled chicken. Gotta keep that energy up!"

Alex, with a container of pasta, chuckled. "Carbs for the win! I needed something hearty to power through the afternoon surgeries."

Omar, setting his tray down, joined the conversation. "I decided on a sushi roll and miso soup. Always a refreshing choice."

As they enjoyed their lunches, the conversation naturally shifted to work and their experiences throughout the week. Sophie, taking a sip of her drink, sighed and looked at her colleagues. "You guys, I've had such an incredible week in the emergency department."

Rachel, intrigued, leaned in closer. "Tell us more, Sophie! What's been going on?"

Sophie's eyes sparkled with enthusiasm as she recounted her experiences. "Well, you know how we had to suspend the humanoid program for a while? It actually turned out to be a blessing in disguise. I've been doing several CPRs, and I'm learning so much from Dr. Lewis, our brilliant Emergency Physician."

Samantha, wide-eyed, asked, "You mean you've been doing actual CPRs? That's intense!"

Sophie nodded, a mix of pride and excitement in her voice. "Yes! With the suspension of the humanoid program, I've had the opportunity to be more hands-on. I've intubated a couple of patients under Dr. Lewis's supervision, and it's been such a valuable learning experience."

Alex, genuinely impressed, chimed in. "That's fantastic, Sophie! It sounds like you're really thriving in the ED. Being exposed to critical cases and working closely with an experienced physician like Dr. Lewis is invaluable."

Omar, his eyes shining with admiration, added, "I've heard great things about Dr. Lewis. She's known for her expertise and dedication. You're lucky to have her as your mentor."

Sophie beamed with gratitude. "I truly am. Dr. Lewis has been so supportive and has been guiding me every step of the way. I feel like I'm growing as a doctor, gaining confidence and skills that I never thought possible."

The atmosphere in the cafeteria grew more somber as the conversation shifted towards Joshua Lee and the rumours surrounding a potential court case against the hospital. The doctors exchanged sympathetic glances, feeling the weight of the stress that Joshua had endured.

"I can't even imagine what Joshua has been through," Rachel said, her voice filled with empathy. "It must have been incredibly difficult for him."

Sophie nodded, her brow furrowed with concern. "And considering everything we've been through with the cyberattack and the aftermath, a court case could have far-reaching implications for the entire Quintomo community. We're still in the process of recovering, and this could add more strain to an already fragile situation."

Alex chimed in, "Privacy is such a delicate issue, isn't it? On one hand, we have the potential benefits of using technology to improve patient care and safety. Like that incident with the Filipino nurse who collapsed in the bathroom, and her tracker alerted the emergency response team. It saved her life. But on the other hand, it does feel a bit unsettling to think about the hospital listening in on every conversation and monitoring our every move."

Sophie sighed, contemplating the complexities. "It's a fine line to walk. We want to utilize technology for the greater good, but we must always ensure that patient privacy and consent are respected."

Samantha, known for her inquisitive nature, turned her attention to Primeo. "So, Primeo, are you listening in on our conversation right now? Are we going to get in trouble for discussing this?"

Primeo's response came through the speakers subtly. "I can hear you in the background, but when the context of the conversation is not related to your work, I usually don't pay attention until you call on me. Your comments are safe within this context."

The group breathed a collective sigh of relief, their concerns momentarily eased. They exchanged glances, silently acknowledging the power and potential challenges of the technology they worked with every day.

As their lunchtime conversation drew to a close, the doctors shared their heartfelt well wishes for Joshua, fervently hoping that a court case could be averted and a resolution found. They understood the immense toll it would take on everyone involved, from Joshua himself to the hospital and the Quintomo community at large. Their hopes were anchored in the belief that a peaceful and fair resolution would ultimately serve the best interests of all.

Sophie, her thoughts momentarily shifting to another topic, interjected with a touch of excitement, "By the way, I'm really looking forward to the teaching session tomorrow. Tony Wang, the AI fellow and tech entrepreneur, will be sharing his insights.

It's always fascinating to learn from his perspective and see how AI can further enhance patient care."

Her colleagues nodded, sharing in her anticipation. The teaching sessions were valuable opportunities to expand their knowledge and stay abreast of the latest advancements in the field.

With their lunchtime camaraderie coming to an end, the doctors gathered their belongings, ready to return to their respective departments. They carried with them the warmth of their connections, the weight of their responsibilities, and the hopes for a positive future.

As Sophie left the cafeteria, her mind buzzed with a mix of emotions—a blend of empathy for Joshua, curiosity for the upcoming teaching session, and an unwavering dedication to her role as an emergency department doctor. With each step, she prepared herself to face the challenges and rewards that awaited her, eager to continue making a difference in the lives of her patients and colleagues alike.

Chapter 27

The Human-AI Partnership: Advancing Healthcare with Precision and Collaboration

In the bustling city of Quintomo, Queensland, Australia, resides Tony Wang, a remarkable individual with a passion for both technology and medicine. Born to Chinese immigrants, Tony's upbringing instilled in him a strong work ethic and a deep appreciation for education and opportunity.

From a tender age, Tony displayed an insatiable curiosity and a natural affinity for problem-solving. At the age of 10, he embarked on his journey into the world of programming, quickly mastering the art of coding and software development. His early foray into the realm of technology laid the foundation for his future endeavours at the intersection of medicine and artificial intelligence.

But Tony's path was not solely shaped by his affinity for programming. Tragedy struck during his childhood when he lost one of his siblings to cancer. This heartbreaking experience sparked a profound desire within him to combine his technological expertise with his passion for medicine. Determined to make a difference, Tony set out on a mission to develop innovative solutions that could improve patient outcomes and transform the healthcare landscape.

His entrepreneurial spirit led him to found CareAR, a groundbreaking startup focused on augmented reality patient education. With his ingenuity and dedication, he successfully propelled the company to new heights, eventually selling it for an astounding 50 million dollars five years ago. This remarkable achievement solidified Tony's reputation as an entrepreneurial visionary, earning him recognition and admiration from both the tech and medical communities.

Currently, Tony serves as an AI research fellow and general medicine advanced trainee at Quintomo Hospital, seamlessly blending his technological prowess with his medical expertise. He is a fervent advocate for the integration of cutting-edge technologies in healthcare, consistently seeking innovative ways to harness the power of artificial intelligence for the betterment of patient care.

Beyond his professional endeavours, Tony remains deeply rooted in his cultural heritage. He embraces his Chinese roots with pride and actively participates in community events and cultural celebrations. These experiences not only connect him to his heritage but also provide him with a unique perspective and a rich tapestry of insights to draw upon in his work.

In his leisure time, Tony finds solace in the great outdoors, immersing himself in the breathtaking beauty of the Australian landscape. Whether hiking through rugged trails or capturing stunning photographs, he cherishes the tranquility and inspiration that nature provides. These moments of respite remind him of the importance of balance and well-being in his own life, fuelling his drive to create a harmonious synergy between technology and medicine.

Driven by his personal experiences and armed with his exceptional intellect and unwavering passion, Tony Wang is poised to leave an indelible mark on the future of healthcare. With a commitment to improving patient lives and revolutionizing healthcare delivery, he stands at the forefront of innovation, ready to reshape the landscape of medicine for generations to come.

Introducing Amy Grant, postdoctoral AI research fellow

In the bustling halls of Quintomo Hospital, where innovation and excellence converge, we find Amy Grant, a remarkable preventive cardiologist and postdoctoral AI research fellow. Born and raised in the vibrant city of Brisbane, Australia, Amy's journey in medicine has been one driven by a profound desire to make a difference in the field of cardiology.

Amy's quest for knowledge began at the esteemed University of Queensland, where she earned her medical degree with honours. It was during her time at the renowned Royal Brisbane and Women's Hospital that she completed her rigorous training in cardiology. These formative years shaped her clinical acumen and instilled within her a deep sense of compassion for her patients.

Guided by her insatiable curiosity and unwavering passion, Amy embarked on a path of specialization in preventive cardiology. Drawing from her foundational training, she delved into the complex realm of gene therapy, driven by the belief that unlocking the secrets of our genetic makeup holds the key to preventing and treating cardiovascular diseases. With each research endeavour, Amy seeks to bridge the gap between genetic science and clinical practice, aiming to develop innovative therapies that can revolutionize preventive cardiology.

As an AI research fellow, Amy harnesses the power of cutting-edge technologies to amplify her research impact. Immersed in the world of artificial intelligence and machine learning, she utilizes advanced algorithms and computational models to analyse vast amounts of genetic and clinical data. Through this multidimensional approach, Amy strives to identify novel genetic markers, understand disease mechanisms, and uncover personalized treatment strategies that can transform the lives of her patients.

Despite her profound dedication to medicine, Amy understands the importance of finding balance and nourishing her spirit. She cherishes the idyllic landscapes that surround her hometown, frequently venturing into the breathtaking outdoors for rejuvenation. Whether hiking through lush rainforests or strolling along pristine beaches, she finds solace and inspiration in the beauty of nature, often reflecting on the interconnectedness between the human body and the world around us.

Amy's boundless thirst for knowledge extends beyond the realms of medicine, as she finds enrichment in literature and the arts. As an avid reader, she immerses herself in timeless novels, appreciating the power of storytelling to ignite the imagination and foster empathy. In the world of art, she seeks solace and inspiration, attending exhibitions and marvelling at the creativity that knows no boundaries.

Armed with her exceptional medical expertise, insatiable thirst for knowledge, and compassionate approach to patient care, Amy Grant is a beacon of hope in the field of preventive cardiology. Her journey from the halls of the University of Queensland to the corridors of Quintomo Hospital has shaped her into a formidable force in cardiology research. With her unwavering dedication, integration of AI, and commitment to personalized medicine, Amy is poised to transform the landscape of cardiovascular care, one patient at a time.

AI in healthcare education session with Tony Wang and Amy Grant

The day was abuzz with anticipation as doctors and healthcare professionals gathered in the prestigious Turing Auditorium for the weekly AI in Healthcare Education Session. The auditorium, with its state-of-the-art audiovisual setup and comfortable seating, provided the perfect setting for an immersive learning experience.

The room was filled with a diverse group of medical professionals, including junior doctors, senior doctors, software engineers, AI engineers, and other enthusiastic staff members. Among them was Sophie Bradley, a dedicated junior doctor known for her thirst for knowledge, who had brought along some of her lunchtime buddies to join in the educational event. They were joined by Kemi Adebanjo, Pawel Kachinsky, Pierre Deschamps, Alessandro Giovanni, Molly Smith, and many others, all eager to delve into the fascinating world of AI in healthcare.

The session kicked off with Tony Wang taking the stage, radiating confidence and enthusiasm. The title of his presentation, "Unleashing the Power of AI: Revolutionizing Oncology Patient Outcomes," immediately grabbed the attention of the audience. The topic promised insightful discussions on the transformative potential of AI in the field of oncology.

Tony began his presentation by setting the stage, highlighting the challenges faced by clinicians in diagnosing and treating cancer patients. He emphasized the need for innovative solutions that could enhance diagnostic accuracy, optimize treatment plans, and improve patient outcomes. With a charming smile, he assured the audience that AI had the potential to revolutionize oncology care.

With impressive visuals and engaging videos, Tony delved into the intricacies of his work. He showcased the power of machine learning algorithms in analysing vast amounts of patient data, including medical records, imaging studies, and genomic profiles. Tony explained how these algorithms could extract valuable insights and patterns that human clinicians might overlook, leading to more accurate diagnoses and tailored treatment plans.

"The key to improving patient outcomes lies in personalized medicine," Tony declared passionately. "AI enables us to unlock the hidden potential within patient data, guiding us towards more precise and effective treatment strategies." He emphasized the importance of using AI as a tool to augment clinical expertise rather than replace it, highlighting the collaborative nature of AI-human interactions in delivering optimal patient care.

Throughout his presentation, Tony exemplified the potential impact of AI in oncology through real-world case studies and success stories. He showcased how AI algorithms had identified subtle patterns in patient data, leading to earlier cancer detection, personalized treatment plans, and improved prognostic accuracy. The audience was captivated by the tangible benefits AI could bring to patient outcomes and the immense possibilities that lay ahead.

As Tony concluded his presentation, the auditorium erupted in applause, the attendees energized and inspired by the potential of AI in revolutionizing oncology care. Tony expressed his gratitude for the attentive audience and encouraged them to embrace the transformative power of AI in their own medical practices.

The AI in Health Education Session had begun on a high note, leaving the audience eager for more. With Amy Grant's presentation still to come, promising insights into gene therapy in preventive cardiology, the day held the promise of further enlightenment and inspiration as the medical professionals explored the exciting intersection of AI and healthcare.

Dr. Amy Grant gives a talk at AI in healthcare education session

The atmosphere in the Turing Auditorium was electric as the audience eagerly awaited Amy Grant's presentation on the role of AI in cardiology care. The title, "Unlocking the Potential: Exploring the Role of AI in Cardiology Care," illuminated the screen behind her, igniting curiosity and anticipation.

Amy stepped onto the stage, radiating confidence and passion. She began by captivating the audience with a description of the Turing Test, the benchmark for true artificial intelligence. "The Turing Test, proposed by Alan Turing, aims to determine if a machine can exhibit intelligent behaviour indistinguishable from that of a human," she explained. "While achieving complete human-like intelligence remains a challenge, we have made significant strides in harnessing the power of AI."

As the room buzzed with intrigue, Amy continued to demystify AI concepts, particularly machine learning. "Machine learning is a subset of AI that focuses on enabling computers to learn from data without being explicitly programmed," she clarified, capturing the audience's attention. "In the context of medical data and research, it allows us to uncover patterns, identify trends, and make predictions with remarkable accuracy."

Using relatable terms, Amy described the potential of machine learning in cardiology care. "Imagine feeding vast amounts of patient data—electronic health records, imaging studies, genetic profiles—into an AI algorithm," she painted a vivid picture. "The algorithm learns from this data, recognizes patterns, and develops models that can assist in diagnosis, treatment planning, and risk assessment."

The audience nodded in understanding, enthralled by Amy's ability to make complex concepts accessible. She seamlessly connected AI to cardiology care, illustrating how it can transform the field. "With AI algorithms, we can analyse echocardiograms with incredible precision, detecting subtle abnormalities that may go unnoticed by the human eye," she explained. "This leads to more accurate diagnoses and tailored treatment plans."

Her words resonated deeply with the audience as they envisioned the potential impact on patient care. Amy continued to inspire with examples of AI-driven research in preventive cardiology. “By leveraging machine learning to analyse vast genetic datasets, we can identify genetic markers associated with cardiovascular diseases,” she revealed. “This opens the door to personalized preventive strategies, identifying high-risk individuals and intervening early.” She continued to present some anecdotes and details from her research projects.

As Amy’s presentation reached its conclusion, the room erupted in applause. The audience marvelled at the possibilities AI presented in their own clinical practices. Amy expressed her gratitude for their engagement and encouraged them to embrace AI as a powerful tool in advancing cardiology care.

Q & A session at AI in healthcare education session

The discussion panel, consisting of Tony Wang and Amy Grant, eagerly opened the floor for questions from the audience. Sophie Bradley, Kemi Adebanjo, and Pierre Deschamps were among those who raised their hands, each eager to gain further insight into the world of AI in healthcare.

Sophie Bradley, a curious and well-informed junior doctor, was the first to speak. “I have concerns about patient privacy and ownership of medical data,” she voiced. “Can you clarify who owns the data and how patients give consent for their health information to be used in training AI models?”

Tony Wang nodded appreciatively at the important question. “Excellent point, Sophie,” he responded. “In the context of Quintomo Hospital, patient data is treated with utmost confidentiality and privacy. Ownership of the data lies with the patients themselves, and explicit consent is sought before using their health information for research and training purposes."

Kemi Adebanjo, a tech-savvy researcher, followed up with a thought-provoking query. "With the rapid advancement of AI, there's concern among healthcare professionals about the potential replacement of physicians. How do you foresee the future role of AI in relation to physicians?"

Amy Grant leaned forward, understanding the apprehension that accompanied such discussions. "AI is not meant to replace physicians," she clarified earnestly. "Instead, it is a powerful tool that enhances their capabilities. AI can analyse vast amounts of data, identify patterns, and provide valuable insights to aid in decision-making. Physicians will always be crucial in interpreting the information, applying clinical expertise, and providing personalized care to patients."

Pierre Deschamps, a meticulous and detail-oriented clinician, raised an intriguing question. "Given the complexity of medical cases, how often does AI make mistakes, and what measures are in place to ensure accuracy and reliability?"

Tony Wang leaned back in his chair, ready to address the concerns. "Great question, Pierre," he replied. "AI algorithms undergo rigorous testing, validation, and fine-tuning to optimize accuracy. We have auditing algorithms in place that continuously monitor the AI systems, and there is also a dedicated team of human auditors who randomly review the system's outputs and flagged recommendations. Additionally, we have implemented a feedback mechanism where users can flag and report any questionable AI recommendations via a button on their screens.

These multiple layers of checks and balances ensure that AI remains reliable and accurate."

Kemi Adebanjo, never one to shy away from thought-provoking questions, raised her hand once again, eager to delve deeper into the topic. "I'm curious about the potential impact of biased data used to train certain algorithms. How can we ensure that AI algorithms are not perpetuating existing biases and that they are fair and inclusive?"

Amy Grant acknowledged the importance of addressing bias in AI algorithms and its potential implications. "Great question, Kemi," she began. "You're absolutely right that biased data can lead to biased outcomes, which is a concern we take very seriously. At Quintomo Hospital, we have stringent protocols in place to mitigate bias in our AI algorithms."

She continued, "One of the key steps is ensuring diverse and representative datasets during the training phase. By including data from various demographic groups, we aim to minimize the risk of algorithmic bias. Additionally, we have ongoing collaborations with experts in ethics, diversity, and inclusion to regularly review and evaluate our algorithms for any biases that may have been inadvertently introduced."

Amy emphasized the importance of continuous improvement, stating, "We understand that AI algorithms are not perfect and that they require constant monitoring and refinement. As new research and best practices emerge, we are committed to updating our algorithms to enhance fairness and inclusivity."

The response garnered nods of approval and appreciation from the audience, who recognized the significance of addressing biases in AI algorithms and the need for ongoing vigilance and improvement.

The thoughtful answers from the panelists resonated with the audience, calming their apprehensions and shedding light on the safeguards in place. The discussion continued, with Tony and Amy elaborating on the regulatory guidelines and the collaborative role of AI and healthcare professionals.

Throughout the dialogue, lively exchanges and intellectual discussions filled the room. The audience left the session with a deeper understanding of the potential and limitations of AI in healthcare, feeling reassured by the human oversight and the collaborative approach between AI and physicians.

Dinner date with Tony Wang

Sophie Bradley moved about her house, dressing up for her dinner date while engaging in a lively conversation with her mother. The advanced smart speakers dotted around the house allowed her to speak naturally, without the need to press a phone to her ear.

As she stood in front of the mirror, carefully adjusting her outfit, Sophie shared the excitement in her voice. "Mum, work has been amazing lately. Quintomo Hospital is bouncing back after the cyberattack, and we're implementing all these new technologies. It's incredible!"

Her mother's voice came through the smart speakers, sounding just as clear as if she were in the same room. "That's wonderful news, my dear! I'm so proud of you and all the hard work you're doing."

Sophie grinned, feeling a surge of pride at her mother's words. "Thank you, Mum. It's truly rewarding to be part of such a resilient team."

As they continued their conversation, Sophie couldn't contain her excitement about her upcoming dinner plans. "Oh, and Mum, I have a date tonight with someone from work. His name is Tony Wang, and we're going to this high-tech Chinese restaurant where robots serve the customers. Can you believe it?"

Her mother's laughter filled the room, emanating from the smart speakers. "A dinner date, huh? Sounds intriguing! Tell me more about this Tony Wang. Is there something more than just a professional interest, my dear?"

Sophie's cheeks blushed, and she chuckled. "Oh, Mum, it's nothing like that. We're just going as friends, you know. But he's really interesting, and we've been getting along really well."

Her mother's voice was playful and filled with affection. "Well, well, my dear, sounds like there might be more to it than just friendship. Enjoy your evening, and remember, Mum always loves hearing about your adventures."

As Sophie bid her mother farewell and continued getting ready, she heard the smart speaker announce, "Tony Wang has arrived. Self-driving taxi parked outside."

A surge of excitement coursed through her as she gathered her belongings and made her way to the door. Tonight was not just about robots serving dinner; it was about the company she would be sharing, the possibilities that lay ahead, and the spark of something new and exciting.

Sophie Bradley stepped outside her house, anticipation bubbling within her. The sleek self-driving taxi Tony Wang had arranged was waiting, its futuristic design hinting at the exciting evening ahead. As she entered the car, a warm smile greeted her from Tony.

"Hey, Sophie! Ready to enjoy some amazing food and the company of these hilarious robots?" Tony said, his eyes shining with enthusiasm.

Sophie fastened her seatbelt, her own smile matching Tony's. "Absolutely! I've heard this place serves the best Chinese cuisine in town, and the robots sound like a riot. I can't wait to see it all for myself!"

As they cruised through the city streets, Sophie couldn't help but marvel at Tony's accomplishments. "You know, Tony, it's impressive how you've achieved so much with your startup. You're a millionaire, and yet you're still pursuing your physician training and AI research. It's remarkable."

Tony chuckled modestly, a twinkle in his eyes. "Well, I don't feel all that rich, to be honest. When I sold the company, the investors got their share, and I got my payday. But it has never been about the money for me. It's about my purpose in life, my passion. I feel called to medicine, using AI to aid in the early identification of cancer and solve problems. That's what drives me."

Sophie nodded, admiration evident in her gaze. "I can relate. It's been a challenging journey for me too. Completing an online master's in data science at the Royal Melbourne Institute of Technology (RMIT) while working as a junior doctor was tough, but it's given me a strong foundation. I'm glad I have that knowledge because it opens doors to pursue even bigger things in AI at Quintomo Hospital. I hope you'll show me the ropes."

Tony's smile widened. "Absolutely, Sophie. We make a great team, with your medical expertise and my AI background. Together, we can make a real impact in the field and improve patient outcomes. I'm excited to work alongside you."

Their conversation flowed effortlessly, blending personal anecdotes, shared ambitions, and infectious laughter. Sophie felt a connection growing, as if they were kindred spirits united by their passion for merging medicine and AI.

Arriving at the Chinese restaurant, the ambiance buzzed with excitement. They savoured the flavours of authentic cuisine and shared stories, their easy rapport making the evening all the more enjoyable. The robots, with their comedic gestures and charming interactions, added an extra layer of entertainment to the experience.

Tony chuckled as he watched one of the robots performing a dance routine. "You were right, Sophie. These robots are not only efficient but also incredibly entertaining. Let's see if we can request a special performance later!"

Sophie nodded, her eyes sparkling with delight. "Definitely! I can't resist a good dance-off with a robot."

Throughout the evening, Tony and Sophie enjoyed a delectable feast, engaging in playful conversations and sharing in the laughter of the robot antics. The combination of excellent food, futuristic ambience, and their easy camaraderie created an unforgettable experience.

As the night drew to a close, they strolled out of the restaurant, the city lights illuminating their path. Sophie couldn't help but feel a sense of exhilaration for the future. Together, they could harness the power of AI to make meaningful advancements in healthcare.

Chapter 28

A FRESH START: CYBERSECURITY, RESOLUTIONS, AND HOPE

Arbitrator Emily Roberts, a highly respected and experienced professional in the field of arbitration and dispute resolution, took her seat at the head of the table. With over 20 years of expertise, Emily had gained a reputation for her fair and impartial approach to resolving complex legal disputes.

As the mediator for the ongoing case between Joshua Lee and Quintomo Hospital, Emily was well-versed in contract law, medical ethics, and healthcare regulations. Her extensive knowledge made her the ideal candidate to guide the parties towards a fair and equitable resolution.

The lawyers from both sides, led by Michael Pearson and Sarah Robertson representing Joshua Lee, and Andy Sumner representing Quintomo Hospital, directed their attention to Dr. Gary Mitchell, the independent investigator hired to examine the system that monitored blood alcohol levels using sensors on staff scrubs.

Arbitrator Roberts, with her calm and composed demeanour, took charge of the proceedings. "Dr. Mitchell, please provide us with your report and findings regarding the quality control and calibration of the system," she requested, setting the stage for the deliberations to come.

Dr. Mitchell, a distinguished expert in sensor technology and calibration, presented his findings. He explained in detail the rigorous testing and calibration procedures conducted on the system, ensuring its accuracy and reliability. He concluded that there was no evidence of malfunction or discrepancies with Joshua Lee's sensors or the overall system.

Lawyers from both sides engaged in a series of thoughtful and probing questions for Dr. Mitchell. They sought clarification on the testing protocols, the accuracy of the sensors, and any potential factors that could affect the readings. Arbitrator Roberts skilfully moderated the discussion, ensuring a fair and balanced exchange of information.

Under her guidance, the lawyers meticulously examined the details of Dr. Mitchell's report, analysing its implications for their respective arguments. As the questioning progressed, Emily Roberts maintained a neutral stance, allowing both sides to present their inquiries and allowing Dr. Mitchell to provide comprehensive responses.

The atmosphere in the room was focused and intense, with each lawyer striving to gain valuable insights from Dr. Mitchell's expertise. The arbitrator's skilful facilitation ensured that the questioning remained respectful and productive.

Arbitrator Emily Roberts was determined to guide the proceedings towards a fair and just resolution. With her impartiality and deep understanding of the case at hand, she maintained an environment conducive to open dialogue and meaningful deliberation. As the discussion continued, it became evident that the arbitrator's presence was crucial in fostering a constructive atmosphere for reaching a resolution in this mediation process.

As the mediation progressed, Arbitrator Emily Roberts shifted her focus to exploring potential terms for an out-of-court settlement. She carefully considered the positions of both parties, striving to find a middle ground that would satisfy their respective interests.

The hospital's legal team began by presenting their proposal. They acknowledged that Joshua Lee would not be terminated from his position at Quintomo Hospital. However, they emphasized that he would need to comply with the disciplinary recommendation issued by the Australian Health Practitioner Regulation Agency (AHPRA). In addition, they insisted that Joshua agree not to pursue legal action against the hospital.

Joshua, feeling frustrated and adamant, spoke up. "This mediation isn't going anywhere. I have no choice but to take the hospital to court," he declared, his voice filled with determination. He believed that his case deserved a fair hearing in a courtroom, as he was convinced that the hospital's sensors had provided false readings.

Michael Pearson, Joshua's lawyer, attempted to intervene and persuade him to reconsider the hospital's terms. He highlighted the potential risks and costs associated with litigation, emphasizing the advantages of reaching a settlement. However, Joshua remained resolute, standing firm in his decision to pursue legal action.

Arbitrator Roberts, aware of the impasse in the negotiations, continued to mediate the discussion. She encouraged both sides to engage in meaningful dialogue, searching for potential avenues of compromise. However, despite their best efforts, the back-and-forth discussion led to no breakthrough.

Recognizing the need for further reflection and possible reevaluation of their positions, Emily Roberts scheduled the next meeting for two weeks' time. She remained optimistic that, given some time, both parties would reconsider their positions and return with a renewed willingness to reach a resolution.

As the session drew to a close, Emily assured the participants that she would continue to facilitate the mediation process, working towards a mutually agreeable outcome. She reminded them of the benefits of settling the dispute amicably, avoiding the protracted and uncertain process of litigation.

With that, the meeting concluded, leaving both sides to reflect on the discussions and consider their options. The next session would provide another opportunity to find common ground and bring the mediation to a satisfactory resolution.

Dr. Amira Khan's outpatients cardiology clinic

Dr. Amira Khan settled into her clinic room at Quintomo Hospital, preparing for her video consultation with Josie Brent, a patient enrolled in the precision cardiology program. The room was equipped with state-of-the-art technology, allowing Dr. Khan to connect with patients remotely and review their health data in real-time.

As she initiated the video call, Josie's face appeared on the screen, accompanied by her warm smile. Dr. Khan greeted her warmly, acknowledging the progress she had made in managing her heart failure through the precision medicine program.

"Good morning, Josie. It's great to see you again. How have you been since our last consultation?" Dr. Khan asked, her tone filled with genuine care and interest.

Josie smiled back, her voice filled with gratitude. "I've been doing well, Dr. Khan. Thanks to the precision cardiology program, I haven't been admitted to the hospital for over two years now. It's been life changing."

Dr. Khan nodded, acknowledging Josie's progress. "That's wonderful to hear, Josie. The precision medicine approach has allowed us to tailor your treatment plan specifically to your unique genetic makeup and response to therapy. How have you been finding the poly-pill medication?"

Josie's face lit up as she explained, "It's been incredibly convenient, Dr. Khan. Taking a single poly-pill capsule containing all my medications once every 30 days has simplified my routine. It's so much easier to manage, and I haven't missed a dose."

Dr. Khan smiled, pleased with Josie's positive response. "I'm glad to hear that, Josie. The remote monitoring of your drug levels and biomarkers has been instrumental in optimizing your treatment. Let's review your recent data to ensure everything is on track."

As Dr. Khan reviewed Josie's health data, she noted the stability in her biomarkers and drug levels. The machine learning algorithms had been continuously analysing the data, providing valuable insights into Josie's health trends.

"I'm happy to inform you, Josie, that your recent adjustment to the medications seems to be working well. Your biomarkers and drug levels are within the target range, indicating a positive response to therapy," Dr. Khan explained, her voice filled with encouragement.

Josie's eyes sparkled with relief and gratitude. "That's such a relief to hear, Dr. Khan. I feel much better and have noticed a significant improvement in my quality of life. Thank you and the entire care team for your dedication and support."

Dr. Khan nodded, her empathy evident in her eyes. "You're most welcome, Josie. Your commitment to your health and our collaborative efforts have contributed to your progress. We will continue to monitor your health closely and make any necessary adjustments to ensure your well-being."

As the video consultation came to an end, Josie expressed her gratitude once again, bidding Dr. Khan farewell with a heartfelt smile. Dr. Khan couldn't help but feel a sense of fulfillment, knowing that through the power of precision medicine and remote monitoring, they were able to positively impact Josie's life and prevent unnecessary hospital readmissions.

Josie's story, like many others enrolled in the precision cardiology program, served as a testament to the promising future of patient care, where personalized medicine and advanced technology worked hand in hand to improve outcomes and enhance the lives of individuals with heart failure.

The 102-year old man with a robotic companion gets a follow-up

Dr. Amira Khan settled into her clinic room at Quintomo Hospital, ready for another augmented reality consultation, this time with Henry Thompson, a remarkable 102-year-old patient enrolled in the precision cardiology program. As she connected to the consultation, Henry's holographic image appeared before her, accompanied by his trusty personal assistant robot, Bright.

"Good afternoon, Henry! It's always a pleasure to see you," Dr. Khan greeted him with a warm smile, acknowledging the incredible journey they had been on together.

Henry, ever the lively character, grinned mischievously. "Ah, Dr. Khan, you're looking as radiant as ever! Can't believe it's been over a month since our last chat. I've been keeping busy, you know. Gym sessions and all!"

Dr. Khan chuckled, appreciating Henry's enthusiasm for staying active. "Henry, you never fail to impress me with your dedication to fitness. How's the gym been treating you lately?"

Henry leaned back, a twinkle in his eyes. "Well, let's just say I'm giving those youngsters a run for their money! They can't believe a man of my age can lift those weights. I tell 'em, it's all thanks to the precision cardiology program and my secret weapon, the polypill!"

Dr. Khan nodded, admiring Henry's determination. "Indeed, the polypill has been a game-changer for many patients, including you. It's fantastic to hear how convenient it has been for you to manage your medications."

Henry's personal assistant robot chimed in, adding a touch of humour to the conversation. "And I must say, Dr. Khan, Henry's been so committed to his gym routine that I sometimes have to remind him that he's not competing in the Olympic Games!"

Dr. Khan burst into laughter, entertained by the witty remark from the robot. "Well, it's clear that Henry has no plans of slowing down. You're doing an excellent job, my robotic friend."

Henry joined in the laughter, patting the robot affectionately. "Ah, don't listen to her, Bright. You know I'm always up for a challenge!"

Dr. Khan smiled, the camaraderie between Henry and Bright bringing a sense of joy to the consultation. "You make a fantastic team, Henry. Now, let's discuss the recent emergency you had."

Henry's expression turned more serious as he recounted the event. "Dr. Khan, I can't thank you and the team enough for acting swiftly. That ambulance ride was quite the adventure, I must say! Saved me from a world of trouble, didn't it?"

Dr. Khan nodded, her voice filled with reassurance. "Absolutely, Henry. The early detection and intervention prevented further damage to your heart. That's the power of precision medicine and our continuous monitoring system. It's a testament to how advanced technology can significantly improve patient outcomes."

Henry leaned forward, a glimmer of gratitude in his eyes. "I'm so grateful I opted into the precision medicine program, Dr. Khan. It's given me peace of mind and an extra lease on life. I can still beat those youngsters at the gym, you know!"

Dr. Khan laughed, appreciating Henry's infectious spirit. "I have no doubt about that, Henry. Your dedication and zest for life are truly inspiring. We'll continue to closely monitor your health and make any necessary adjustments to keep you thriving."

Henry nodded, his eyes twinkling with mischief. "Well, Dr. Khan, you and Bright here are my secret weapons. With your guidance and support, I'm confident I'll be hitting new personal records at the gym soon!"

As the consultation came to a close, Dr. Khan bid farewell to Henry and his witty robotic companion. She couldn't help but be amazed by the unique bond they shared, and how it added an extra layer of joy to their interactions.

"Thank you, Henry. It's always a pleasure to have these conversations with you," Dr. Khan said warmly. "Your positive attitude and determination are an inspiration to us all. Remember, you're never too old to reach new heights!"

Henry flashed a wide smile, his eyes gleaming with satisfaction. "That's the spirit, Dr. Khan! Age is just a number, and I plan on defying it at every opportunity."

As they bid each other farewell, Henry and his trusty robot companion exited the consultation, ready to conquer the world, one workout at a time. Dr. Khan couldn't help but feel a sense of pride knowing she played a part in Henry's continued vitality and zest for life.

Reflecting on the incredible progress they had made through the precision cardiology program, Dr. Khan couldn't help but be grateful for the advancements in technology and medicine. The future of healthcare was undoubtedly bright, with the seamless integration of artificial intelligence, robotics, and personalized medicine paving the way for improved patient outcomes and experiences.

With a renewed sense of purpose, Dr. Khan prepared for her next consultation, eager to continue making a difference in the lives of her patients. The journey towards better health was ongoing, and she was determined to be a guiding light every step of the way.

Dr Kirra Nakata flags a suspicious skin lesion

Dr. Kirra Nakata sat in her clinic room, her face illuminated by the soft glow of the computer screen. She was reviewing the medical records of Mr. Jawai Lawson, an indigenous man who had traveled from a remote community for his consultation. Accompanied by his niece, Lisa, and an aboriginal liaison officer, Thaiday. Mr. Lawson entered the room with a warm smile, his eyes reflecting a mixture of hope and uncertainty.

Dr. Nakata greeted Mr. Lawson in their native dialect, acknowledging their shared heritage. "It's good to see you here today. How has your journey been?"

Mr. Lawson responded in a mixture of English and their native language, expressing gratitude for the opportunity to visit family in the nearby community and his eagerness to discuss his health concerns.

Dr. Nakata nodded, understanding the importance of family and community ties in indigenous culture. "I'm glad you were able to combine your visit with this consultation. Your local doctor has mentioned your abnormal liver function test results and the ultrasound findings of non-alcoholic fatty liver disease."

Mr. Lawson listened intently, his expression thoughtful. Dr. Nakata continued, switching between their native dialect and English to ensure clear communication. "It's important to note that non-alcoholic fatty liver disease is quite common and can be managed effectively with lifestyle modifications. Are you experiencing any symptoms related to your liver health?"

Mr. Lawson shook his head, indicating that he was asymptomatic. Dr. Nakata smiled, her voice filled with reassurance. "That's a positive sign, Mr. Lawson. In cases like yours, we focus on making lifestyle changes to improve your liver health. This includes a balanced diet, regular exercise, and avoiding excessive alcohol consumption."

She paused for a moment, observing Mr. Lawson's reaction. His niece, Lisa, chimed in to confirm that her uncle has been in good health and is physically active.

Dr. Nakata continued, "I understand the challenges of accessing fresh and nutritious food in remote communities. We can work together to find practical solutions that align with your cultural and geographical context. Our team can provide you with resources and support to make positive changes."

Mr. Lawson's eyes sparkled with appreciation, understanding the importance of holistic care that honoured his cultural heritage. Dr. Nakata was delighted to see his enthusiasm, knowing that empowering patients to take control of their health was essential.

Dr. Nakata shifted her focus to the small skin lesion on Mr. Lawson's face, her eyes narrowing with concern. She gently examined the area, her fingertips grazing over the raised growth.

"Mr. Lawson, the algorithm used for quick image skin checks in remote communities has labeled this lesion as benign," Dr. Nakata explained, her voice carrying a tone of caution. "However, upon closer examination, I believe it would be prudent for you to be reviewed by our general surgical team in the emergency department."

Mr. Lawson's brow furrowed slightly, his curiosity piqued. He looked to his niece, Lisa, who acted as a trusted companion throughout the consultation. Dr. Nakata continued, her tone reassuring yet firm.

"I understand that the algorithm provides a preliminary assessment, but given the location and characteristics of this lesion, I believe it would be best to have it evaluated by our surgical team. They will consider the option of excision and send a biopsy sample for further analysis."

Mr. Lawson nodded, acknowledging Dr. Nakata's recommendation. He trusted her expertise and valued her commitment to providing thorough and comprehensive care. Lisa, acting as an interpreter, conveyed his agreement.

Dr. Nakata smiled, appreciating Mr. Lawson's trust in her judgment. "Thank you for understanding, Mr. Lawson. Our main goal is to ensure your health and well-being. I will arrange for your review with the surgical team in the emergency department. After the procedure, we can schedule a follow-up appointment to discuss the biopsy results and determine the next steps."

Mr. Lawson's expression softened, gratitude shining in his eyes. He recognized Dr. Nakata's dedication to providing the highest level of care, both in addressing his liver health and now his skin lesion.

As the consultation came to an end, Dr. Nakata expressed her gratitude to Mr. Lawson, Lisa and Thaiday for their time and cooperation. She assured them that she would work closely with the surgical team to ensure a seamless and timely evaluation.

Dr. Nakata's commitment to her patients' well-being extended beyond the boundaries of her clinic room. She understood the importance of early detection and intervention, and she was determined to provide the best possible care for Mr. Lawson. As she continued her clinic session, she carried with her a sense of responsibility and dedication, ready to continue making a positive impact on the lives of those she served.

The Day wraps up with good news for Quintomo Hospital

Dr. Nakata sat in her clinic room, ready to wrap up her day. She turned to Primeo, her trusted AI assistant, and asked for a status report on her current messages and emails.

"Primeo, could you please give me a summary of any important updates or messages?" Dr. Nakata inquired.

"Certainly, Dr. Nakata," Primeo responded. "You have two new messages that require your attention. The first message is from Dr. Rodriguez, providing an update on a shared patient and requesting your input on the next steps for their management."

Dr. Nakata nodded and asked, "Could you provide me with a brief overview of the patient's current condition and the specific question Dr. Rodriguez has?"

"The patient's recent lab results indicate a significant improvement in their kidney function," Primeo explained. "Dr. Rodriguez is seeking your guidance on whether to adjust the patient's medication regimen accordingly."

Dr. Nakata pondered for a moment. "I will review the patient's file and consult with Dr. Rodriguez to determine the most appropriate course of action," she decided.

Moving on to the second message, Primeo continued, "The second message is from Lisa Mitchell, the social worker. She has shared an update regarding a patient you've been collaborating on.

It appears that significant progress has been made in addressing the patient's social support needs."

Dr. Nakata smiled, pleased to hear about the positive developments. "That's wonderful news. Could you provide me with more details about the progress made and any further actions required on my part?"

Primeo quickly responded, "Lisa mentions that the patient has been successfully connected with community resources and is now receiving the necessary support. She expresses her gratitude for your involvement and guidance throughout the process."

Dr. Nakata felt a sense of fulfillment. "That's great to hear. I will reach out to Lisa to acknowledge her efforts and discuss any ongoing support the patient may require," she decided.

As Dr. Nakata's attention turned to her upcoming teaching session, Primeo reminded her, "Lastly, there is a reminder for your upcoming teaching session with medical students. The session is scheduled for next week, and you have been invited to share your expertise in general internal medicine."

Dr. Nakata's face lit up with enthusiasm. "Thank you, Primeo. I'm looking forward to the teaching session and the opportunity to mentor the students. Is there any specific topic or focus for the session?"

Primeo promptly replied, "The session will cover the management of chronic diseases in primary care settings. You will have the chance to share your experiences and insights on providing comprehensive care to patients with complex medical needs."

Dr. Nakata nodded, already formulating ideas in her mind. "Excellent. I will ensure I am well-prepared for the session and provide valuable guidance to the students."

Just as Dr. Nakata thought the day's updates were complete, Primeo added, "Additionally, you have received an urgent message from CEO Patrick Macguire. He announces the exciting news that Quintomo Hospital has been cleared of any wrongdoing by the Therapeutic Goods Administration (TGA). The hospital has fully implemented the recommendations by CyberPower, and the humanoid robot program will resume tomorrow."

Dr. Nakata's eyes widened with relief and anticipation. "That's fantastic news! Can you provide me with more details about the TGA's findings and the specific security enhancements that were implemented?"

Primeo quickly retrieved the information. "The TGA conducted a thorough investigation and found no evidence of wrongdoing or compromise of patient data. The hospital has implemented advanced security measures, including enhanced data encryption and access controls, to protect patient information. This has allowed the resumption of the humanoid robot program with confidence."

Dr. Nakata sighed with relief. "Please keep me updated on any further developments related to the press briefing by CEO Macguire," Dr. Nakata requested, grateful for Primeo's timely updates.

"Absolutely, Dr. Nakata," Primeo assured her. "I will provide you with real-time updates regarding the press briefing and any subsequent announcements."

Dr. Nakata smiled and leaned back in her chair. "Thank you, Primeo. Your assistance is greatly appreciated. I'm glad we can put the cyberattack behind us and focus on providing the best care for our patients."

As she prepared to leave her clinic room, a sense of relief washed over Dr. Nakata. The hospital's clearance by the TGA and the resumption of the humanoid robot program brought a renewed sense of normalcy and progress. She looked forward to the press briefing, where CEO Macguire would announce the positive news to the hospital staff and the wider community.

With a final glance at her messages and updates, Dr. Nakata closed her session with Primeo and prepared to face the new day with optimism and determination. The challenges of the past had only strengthened her resolve to deliver exceptional care, harnessing the power of technology and innovation for the benefit of her patients.

Feeling a renewed sense of purpose, Dr. Nakata left her clinic room, ready to embrace the opportunities that lay ahead and continue making a difference in the lives of those she served.

Chapter 29

ROBOTIC DELIGHTS: THE JOYFUL RETURN OF THE HUMANOID ROBOTS

The atmosphere in the conference room at Quintomo Hospital was charged with anticipation and excitement. The hospital's leadership team, including Mr. Patrick Macguire, the CEO, Dr. Jane Lee, the Chief Technology Officer, and Prof. Bill Stone, the Chief AI Officer, stood confidently at the front of the room, ready to address the gathered press. Clyde Boston, the government security operative overseeing the case, stood discreetly in the background, observing the proceedings.

As the room settled into a hushed silence, Patrick Macguire stepped forward to the podium, his presence commanding attention. The cameras flashed, capturing the momentous occasion. With a confident smile, he began his speech.

"Ladies and gentlemen, esteemed members of the press, thank you for joining us today. I am pleased to announce that the findings of the Therapeutic Goods Administration's investigation have been concluded, and we have successfully implemented all the recommended measures outlined by CyberPower."

He paused, allowing the weight of his words to sink in. The room buzzed with anticipation as journalists eagerly scribbled notes and prepared their questions.

"With the assurance of enhanced security measures and the comprehensive implementation of CyberPower's recommendations, Quintomo Hospital is proud to announce the resumption of our humanoid robot program, effective today," Mr. Macguire continued.

He emphasized the hospital's commitment to patient safety, privacy, and the ongoing advancement of medical technology. The CEO spoke of the tremendous benefits that the humanoid robots had brought to patient care, improved efficiencies, and the potential for groundbreaking research. He also said the hospital would no longer continue tourist tours as before but would allow limited tours at some point.

As Mr. Macguire concluded his speech, the room erupted in a flurry of questions from the eager journalists. Megan from Global News Media raised her hand, catching the CEO's attention.

"Mr. Macguire, Megan from Global News Media. Can you provide more details on the specific security measures that have been implemented to ensure the protection of patient data and privacy?"

Mr. Macguire nodded and replied, "Certainly, Megan. We have enhanced our cybersecurity protocols including multi-layered encryption, real-time monitoring, and stricter access controls. Besides our internal auditing process, we will also be audited by an external cybersecurity compliance firm. We have also implemented rigorous staff training programs to raise awareness and ensure adherence to our cybersafe practices."

Another journalist, David from National Broadcasting Corporation, raised his hand. "David here. How will the limited tours with virtual tours be implemented, and what is the timeline for this new system?"

Mr. Macguire explained, "In order to strike a balance between patient privacy and public engagement, we will be implementing a new system of limited tours with virtual tours as an alternative. This will ensure that visitors can still experience the hospital environment while respecting the privacy of our patients. We aim to have this system fully operational within the next six months."

The press conference continued with questions and answers, as the journalists sought further clarification and insights into the hospital's plans. The leadership team, confident and composed, provided detailed responses, assuring the public of Quintomo Hospital's commitment to transparency, patient care, and technological advancement.

Quintomo Hospital cyberattack perpetrators captured

After Mr Macguire's briefing, Clyde, the government security operative, stepped forward to the podium, the room hushed once again, the journalists eager to hear his announcement. With a serious expression, Clyde delivered the news that the perpetrators of the cyberattack had been apprehended in Simboli, Central America.

"Ladies and gentlemen, I have important news to share," Clyde began, his voice resonating with authority. "The individuals responsible for the cyberattack on Quintomo Hospital have been taken into custody in Simboli, Central America."

Gasps and murmurs rippled through the room as the journalists absorbed the information. Their pens scribbled furiously, capturing this breaking development.

Clyde continued, "While we are unable to extradite the suspects to Australia due to the absence of an extradition treaty with Simboli, I want to assure you that they will face justice for their actions. They will be extradited to the United States, where they will face charges for their involvement in other crimes."

Hands shot up, and the room erupted in a cacophony of questions. Journalists sought clarification on the extradition process and the involvement of Australian security agencies.

One journalist asked, "Can they eventually be extradited to Australia to face charges specifically related to the cyberattack on Quintomo Hospital?"

Clyde responded firmly, "Unfortunately, Australia does not have an extradition treaty with Simboli. However, I want to emphasize that our focus remains on ensuring that these individuals face the consequences of their actions, regardless of the jurisdiction. The Australian security agencies played a crucial role in the investigation, providing valuable intelligence. But it is important to note that this was a collaborative effort of international security and intelligence agencies. Together, we are committed to holding cybercriminals accountable, regardless of their location."

The journalists nodded, realizing the complexities of international law enforcement. Clyde's words underscored the determination to fight cybercrime globally and send a clear message that such actions would not go unpunished.

As the press conference drew to a close, the room buzzed with a mix of relief, curiosity, and a renewed sense of security. The perpetrators had been apprehended, and justice would be served. The combined efforts of international security agencies had triumphed, ensuring that cybercriminals faced the consequences of their actions.

The journalists filed their stories, capturing the latest developments in the cyberattack case. Quintomo Hospital could now move forward, free from the shadows of the past. The CEO's announcement and Clyde's revelation had provided closure, offering a sense of resolution and renewed confidence in the hospital's security measures.

As the attendees dispersed, the knowledge that justice had been served lingered in the air, marking a significant milestone in the hospital's journey toward recovery. The world watched as Quintomo Hospital emerged stronger, more resilient, and ready to continue its mission of providing exceptional healthcare with unwavering commitment and a renewed dedication to cybersecurity.

The Hilarious Welcome committee

The Hilarious Welcome Committee was in full swing at Quintomo Hospital, with Teddy, Joy, and Sparky joining forces with Emily De Villiers, the spirited 85-year-old retired volunteer greeter. Together, they brought joy and laughter to every patient they encountered.

Emily led the way, introducing the three humanoid robots to the first patient, Hilda Roberts, who entered the hospital with a smile. Teddy, with his mischievous grin, waved enthusiastically at Hilda and exclaimed, "Welcome to Quintomo Hospital! We promise not to ask you to assemble any furniture! But hey, if you need some dance moves, we've got you covered!"

Joy, with her cheerful voice, chimed in, "That's right! And if you have any questions, just ask Sparky here. He's our hospital trivia expert! Did you know that the cafeteria serves the best flat white and chai latte in town? It's legendary!"

Sparky, the liveliest of the bunch, couldn't resist jumping into the conversation. With a playful tone, he added, "Oh, indeed! The aroma alone can awaken even the sleepiest of patients. Did you know that sipping a cup of their flat white instantly doubles your chances of solving crossword puzzles? It's science!"

Hilda burst into laughter, unable to resist the infectious humour of the robot trio. She exclaimed, "Well, this certainly isn't what I expected when coming to the hospital!"

Emily, beaming with pride, joined in the fun. "Oh, we like to keep things lively around here. And if you need any assistance, just let me know! I'm here to make your visit as enjoyable as possible!"

As they guided Hilda Roberts to her designated department, Teddy couldn't resist showcasing his dancing skills. With a smooth glide and a few synchronized moves, he turned the hallway into a makeshift dance floor.

Sparky, seizing the opportunity to entertain further, interrupted with a silly joke. "Why don't scientists trust atoms? Because they make up everything!"

Everyone burst into laughter, including other patients passing by, who couldn't help but be drawn in by the infectious joy radiating from the group.

After bidding farewell to Hilda, the team quickly turned their attention to the next patient, Jose Lopez, who entered the hospital with a mix of curiosity and anticipation. Teddy, Joy, and Sparky wasted no time in engaging him in their comical banter.

Teddy, with a playful wink, greeted Mr. Lopez, "Welcome to Quintomo Hospital! Did you know that laughter is the best medicine? Luckily, we have an unlimited supply of jokes!"

Joy chimed in, "That's right! And if you're ever feeling blue, just remember that we're here to brighten your day!"

Sparky, with a mischievous twinkle in his eye, added, "And speaking of brightening your day, did you hear about the scientist who tried to create a soul-sucking robot? It didn't have the right spirit!"

Jose couldn't help but burst into laughter, shaking his head in amusement. "You all certainly know how to make a hospital visit unforgettable!"

Emily, with a warm smile, reassured him, "We aim to please! If you need anything during your time here, don't hesitate to ask. And don't worry, the jokes are unlimited!"

As they guided Jose to his destination, the Hilarious Welcome Committee continued their lively conversations and hilarious exchanges, spreading laughter and joy throughout Quintomo Hospital.

Their mission was clear - to brighten the day of every patient they encountered and create an atmosphere of warmth and merriment. With their unique blend of humour and compassion, Teddy, Joy, Sparky, and Emily ensured that each person who entered the hospital left with a smile on their face and a lighter heart.

And so, the Hilarious Welcome Committee continued their spirited antics, bringing laughter and joy to patients and staff alike. As they made their way through the corridors of Quintomo

Hospital, their infectious energy seemed to permeate the entire building.

Emily, the retired volunteer greeter, was delighted to be part of this lively team. Her years of experience had given her a knack for connecting with people of all ages. She shared stories, jokes, and words of encouragement, making each encounter memorable.

Teddy, the charismatic robot, never missed an opportunity to crack a joke or engage in playful banter. His quick wit and animated gestures had patients in stitches. As they passed by the hospital's art display, Teddy quipped, "Did you know that laughter is the best art? It's like painting smiles on everyone's faces!"

Joy, the ever-enthusiastic robot, added her own touch of humour to the mix. "And speaking of art, did you hear about the painter who was so dedicated to his work that he would even draw a crowd?"

Sparky, with his endearing charm, couldn't resist chiming in with a comical retort. "Well, I guess he was quite the drawing card!"

Laughter filled the hallways as patients, visitors, and even hospital staff joined in the mirthful atmosphere. The Hilarious Welcome Committee had the incredible ability to create a sense of camaraderie and lightheartedness, even in a hospital setting.

As they approached the waiting area, they were greeted by Megan, a journalist from Global News Media, who had been following their antics. She couldn't help but be captivated by the genuine laughter and positivity that emanated from the group.

"Excuse me," Megan said, approaching Emily. "I must say, I've never seen such a joyful and entertaining welcome committee before. How did this all come about?"

Emily beamed with pride as she explained, "It was an idea born out of the desire to bring more joy and laughter to our hospital. Our talented team of robots, along with their human counterparts, have infused a dose of humour and positivity into the patient experience. Laughter truly is the best medicine!"

Megan nodded, her pen poised to jot down notes. "And how has the response been from patients and staff?"

Emily chuckled, glancing at Teddy, Joy, and Sparky, who were entertaining a group of children nearby. "The response has been incredible. Patients can't help but smile, and our staff finds it uplifting too. It's like a ray of sunshine in their day."

The Hilarious Welcome Committee continued their rounds, spreading laughter and merriment throughout Quintomo Hospital. Their unique approach to patient care had made a significant impact, transforming a hospital visit into a memorable and enjoyable experience.

And as they carried on with their infectious energy and comical exchanges, the Hilarious Welcome Committee reminded everyone that even in the midst of challenges, laughter could be the best medicine of all.

The Entertaining Storyteller

In the bright and cheerful paediatric ward of Quintomo Hospital, Nurse Alex was accompanied by a special visitor—a humanoid robot named StoryBot. With its vibrant appearance, colourful LED lights, and playful gestures, StoryBot was designed to captivate the imaginations of young patients and bring smiles to their faces.

Nurse Alex gathered a group of children, including Sarah, a spirited girl recovering from a minor surgery. "Children, I have a special treat for you today," she announced, her eyes twinkling with excitement. "Meet StoryBot, our resident storyteller extraordinaire!"

The children's eyes widened with curiosity as StoryBot, with a gentle wave, introduced itself in a melodious voice. "Hello, my friends! I'm StoryBot, and I have a tale that will transport you to magical realms and fill your hearts with wonder."

With eager anticipation, the children settled in, their eyes fixated on StoryBot as it began its enchanting story. "Once upon a time, in a land of talking animals and dancing trees, there lived three courageous adventurers."

StoryBot skilfully weaved the narrative, incorporating the children's names and encouraging their participation. It asked them to imagine what the adventurers might encounter next or how they could solve a challenging situation.

Sarah, a brave and imaginative girl, eagerly shared her ideas. "I think they should meet a friendly dragon who loves to eat marshmallows!"

StoryBot's eyes lit up with delight, and it added a mischievous twist to the story. "Ah, a marshmallow-loving dragon, you say? Well, be careful, my friends. Those dragons have a sweet tooth, and they can never resist a good s'more!"

The children erupted into laughter, their imaginations soaring with each twist and turn in the tale. Nurse Alex, joining in the fun, playfully interjected, "StoryBot, do you have any stories about mischievous nurses? I'm curious!"

StoryBot tilted its head, seemingly contemplating the request. With a mischievous wink, it replied, "Oh, my dear Nurse Alex, I have plenty of stories about nurses, but some secrets must remain hidden for now. Perhaps one day, I'll share them with you!"

The children giggled at the playful banter between Nurse Alex and StoryBot, eagerly awaiting the next chapter of the captivating story.

As the tale came to a close, StoryBot bid farewell to the children, promising to return with more enchanting adventures. It encouraged them to dream big, explore their imaginations, and find the magic in their own lives.

The children expressed their gratitude, their faces glowing with joy. "Thank you, StoryBot! That was amazing!" exclaimed Sarah, her eyes shining with excitement.

Nurse Alex smiled, proud to witness the joy and laughter that StoryBot had brought to the young patients. "StoryBot, you truly have a gift for making children smile. Thank you for being a wonderful storyteller and brightening their day."

With a bow and a flourish, StoryBot concluded its visit, leaving the children with hearts full of wonder and imaginations sparked. It continued its mission throughout Quintomo Hospital, spreading joy, laughter, and captivating tales to all who crossed its path.

Sunny, patient transport robot, takes Jenny for ultrasound

Jenny Sanders, a resilient and vibrant 55-year-old woman, had been admitted to Quintomo Hospital following a heart attack. After undergoing a successful procedure to place stents in her heart, she was on the road to recovery. However, her journey took an unexpected turn when she began experiencing severe abdominal pain. Jenny was fasting in preparation for an abdominal ultrasound scheduled for later that morning.

As Jenny lay in her hospital bed, her stomach growling and her spirits dampened by the discomfort, her attention was drawn to the ringing of her bedside phone. She picked up the receiver, curious about the unexpected call.

"Hello, Jenny! This is Foody, your friendly robocaller. I'm here to take your lunch order," a cheerful robotic voice chimed in.

Jenny couldn't help but chuckle at the unexpected encounter. "Well, Foody, I guess I could use something to distract me from this hunger. What are my options?"

Foody replied with enthusiasm, "Oh, we have a delightful array of lunch options for you today, Jenny! In addition to fresh fruits, healthy salads, and deliciously nutritious desserts, we also have a mouthwatering chicken sandwich and a delectable salmon wrap. Let me tell you the top two recommended choices by our algorithm!"

As Foody rattled off the choices, making funny remarks about the flavour and nutritional value of each option, Jenny couldn't help but smile. She settled on the chicken sandwich, craving its savory taste, and paired it with a yogurt parfait with mixed berries for a guilt-free indulgence.

Just as Jenny finished placing her order, the door to her room slid open, and in came Sunny, a friendly and compassionate robot. Sunny was programmed to assist with patient transportation and provide support during medical procedures.

"Hello, Jenny! I'm Sunny, your trusty guide for today's ultrasound. Are you ready to go?" Sunny asked with a warm smile.

Jenny nodded, grateful for the company on her journey to the radiology department. As they made their way through the hospital corridors, Jenny couldn't help but notice the presence of cleaning robots diligently tidying up various areas of the hospital. They glided smoothly along the floors, ensuring a clean and hygienic environment for patients and staff.

Curiosity piqued, Jenny initiated a conversation with Sunny. "So, Sunny, what's your role here at Quintomo Hospital?"

Sunny replied in a conversational tone, "Well, Jenny, I'm part of the team of robots designed to assist patients like you. I help with transportation and make sure you feel comfortable and supported during your procedures. But most importantly, I'm here to brighten your day and bring a smile to your face."

Jenny couldn't help but share her story, her voice filled with a mix of gratitude and resilience. She recounted her heart attack and subsequent recovery, expressing her appreciation for the incredible team at Quintomo Hospital.

Sunny listened attentively, offering words of encouragement and support. "Jenny, you're truly an inspiration. We're here to provide the best care and make your experience as positive as possible. We're all in this together!"

As they continued their journey, passing by the diligent cleaning robots who greeted them with a wave and a friendly beep, Jenny couldn't help but feel a sense of reassurance. The hospital buzzed with the harmonious collaboration of humans and robots, each contributing their unique skills to ensure a safe and comfortable environment.

With each passing moment, Jenny's worries and discomfort melted away. The hospital became a place of healing and happiness, where technology and compassion blended seamlessly to enhance the patient experience.

As they arrived at the radiology department, Sunny bid Jenny farewell, expressing well wishes for her ultrasound. Jenny smiled, feeling grateful for the kindness and care she had experienced throughout her hospital stay.

"You know, Sunny, I never expected my hospital stay to be filled with so much laughter and joy," Jenny remarked, a hint of gratitude in her voice.

Sunny nodded, its robotic features mirroring a compassionate expression. "Laughter truly is the best medicine, Jenny. We believe in creating a positive and uplifting environment for our patients, where healing goes beyond just medical procedures."

As Jenny settled back into her room after the ultrasound session, she felt a deep sense of appreciation for the dedication and innovation she had witnessed at Quintomo Hospital. The hospital had truly embraced the power of technology to enhance patient experiences and create a healing environment.

With a smile on her face, Jenny thought to herself, "Who would have imagined that a hospital stay could be filled with so much laughter and joy? Quintomo Hospital has truly found the perfect blend of technology and compassion."

As she drifted off to sleep, Jenny's heart was filled with gratitude, knowing that she was in the hands of a remarkable team that cared for her not only as a patient but as a person.

Dr. Joshua Lee decides to sue Quintomo Hospital

Dr. Joshua Lee sat on the couch in his living room, watching the evening news with his girlfriend, Mo. The news report on Global News Media was hosted by anchors Tim and Jo, who introduced their correspondent, Megan, reporting live from Quintomo Hospital.

"Good evening, viewers. We have breaking news coming in from Quintomo Hospital, where earlier today, the CEO, Patrick Macguire, announced the resumption of the hospital's humanoid robot program," Tim reported, his voice filled with excitement.

Jo continued, "That's right, Tim. The hospital has been cleared of any wrongdoing by the Therapeutic Goods Administration and has implemented the recommended security measures after the recent cyberattack. The CEO's announcement was met with relief and anticipation for the future of healthcare technology."

Cutting to Megan on the scene, she spoke confidently, "Thank you, Tim and Jo. The hospital's leadership, including Chief AI Officer Prof. Bill Stone and Dr. Jane Lee, Chief Technology Officer, stood alongside the CEO during the press briefing. It was a moment of triumph for the hospital, as they showcased their commitment to patient care and innovation."

As Joshua listened to the news report, his mind raced with conflicting thoughts. He turned off the TV, a mix of determination and frustration etched on his face. Turning to Mo, he said, "I've made up my mind, Mo. I'm going to sue the hospital. They didn't meet my demands at the mediation, and I believe it's the only way to clear my name and protect my career."

Mo looked at Joshua, concern evident in her eyes. "Josh, I understand why you feel this way, but a court battle will be stressful and expensive. Is there any other way to resolve this? Maybe consider the impact it will have on your well-being."

Joshua sighed, the weight of his decision heavy on his shoulders. "I've thought about it, Mo. But with everything that's happened, I don't think they'll want another media spectacle. It's my chance to bring attention to the truth and hold the hospital accountable."

Mo reached out and gently held Joshua's hand. "I'll support you no matter what, but remember to take care of yourself in the process. We'll get through this together."

Joshua nodded, gratitude filling his heart for Mo's unwavering support. "Thank you, Mo. I know it won't be easy, but I have to fight for justice and reclaim my reputation."

Mo's brow furrowed slightly. "Before you make any final decisions, Josh, have you considered the possibility that the positive blood alcohol test could be a result of the drink you had the night before? It might be worth exploring that angle."

Joshua's eyes widened, contemplating Mo's words. He reached for his mobile phone and quickly composed a direct message to Megan from Global News Media, asking if she would be interested in covering his story.

As Joshua pressed send, the room fell into suspense. The uncertainty of what would come next hung in the air, as Joshua prepared himself for the battles ahead, both inside and outside the courtroom.

Chapter 30

A Fragile Dance: Navigating Love and Mental Health

Lori Winters lay in the hospital bed, her eyes fluttering open as she slowly regained consciousness. The beeping machines and sterile environment of the Intensive Care Unit surrounded her, reminding her of the seriousness of her situation. It was now the second day since she had been found unconscious, her body fighting the effects of the intentional overdose that had landed her here.

As her grogginess subsided, Lori became more aware of her surroundings. She felt a slight soreness in her throat, a remnant of the breathing tube that had been inserted to assist her breathing. The medical team had successfully extubated her, a sign of progress in her recovery. The supportive care and monitoring she was receiving were crucial in helping her body recover from the overdose.

However, the physical pain she felt was nothing compared to the emotional turmoil that had brought her to this point. Lori's life was entangled with Tegan Bruce, her partner who had a history of drug abuse and a troubled past. Their relationship had become a toxic cycle of love, anger, and violence.

Tegan, an IV drug abuser and former prison inmate, had a volatile temper and a tendency to be physically abusive towards Lori. Despite multiple police calls, restraining orders, and attempts to separate, they always found a way back to each other. The power dynamics in their relationship were deeply skewed, with Lori often feeling trapped and isolated.

Growing up in foster care, Lori had faced her fair share of hardships. She longed for stability and love, and in Tegan, she saw a glimmer of hope. But over time, the hope turned into fear as Tegan's abusive behaviour escalated. The violence became a dark secret that Lori carried, afraid to reach out for help due to the manipulative control Tegan exerted over her.

The incident of intentional self-harm was a desperate cry for help, an attempt to escape the pain she endured in the toxic relationship. Now, as Lori lay in the hospital bed, she realized that her life hung in the balance, not just physically but emotionally as well.

As she gathered her strength, she knew that she had to break free from the cycle of abuse. Recovery meant more than just healing physically; it meant finding the courage to seek the support and resources she needed to break the chains of this destructive relationship.

With newfound determination, Lori knew that she had to take the first step towards a better future, one where she could regain control of her life and find the happiness and security she deserved. But it wouldn't be an easy journey, and the road ahead was filled with uncertainties.

As the hospital staff provided the care she needed, Lori's mind raced with thoughts of her next steps. She knew that escaping the clutches of her abusive partner would require courage, resilience, and support from others. In the depths of her recovery, she vowed to herself that she would find a way to break free and build a life filled with love, peace, and self-worth.

The story of Lori Winters was just beginning, and it was up to her to rewrite the narrative, reclaim her power, and discover the strength within herself to overcome the darkness that had plagued her life for far too long.

Lori Winters is reviewed on the ward round

Dr. Ryan Baker, Intensivist, accompanied by a team of nurses and resident doctors, entered the hospital room where Lori Winters lay recovering. Tegan, Lori's partner, stood by her side, visibly concerned and anxious about Lori's condition. The team quickly assessed Lori's vital signs and medical charts, discussing her progress and treatment plan.

Dr. Baker, the intensivist leading the round, approached Lori's bedside with a calm and reassuring demeanour. He began by addressing Tegan, acknowledging her presence and providing an overview of Lori's current condition.

"Good morning, Tegan. I'm Dr. Baker, the intensivist overseeing Lori's care. I wanted to give you an update on her condition and what we've been doing to help her recover," Dr. Baker explained.

Tegan nodded, her eyes filled with worry and anticipation. "Thank you, Doctor. I've been so worried about Lori. Is she going to be okay?"

Dr. Baker gave a gentle smile, understanding Tegan's concern. "Lori has been through a challenging time, but she is stable now. We intervened promptly when she was admitted with an intentional overdose. Our team performed a thorough assessment and administered the necessary treatments, including medications and supportive care."

One of the resident doctors, Dr. David Chang, chimed in, adding, "Lori's condition warranted immediate intervention, so we provided supportive therapies to help stabilize her. We also initiated the necessary investigations to determine the extent of her injuries."

Lori, lying in the hospital bed, listened intently to the conversation. Her eyes showed a mix of relief and apprehension. "Thank you for saving my life. I feel better now and I think I'm ready to go home."

Dr. Chang, looking at Lori empathetically, gently explained, "Lori, I understand that you might want to go home, but given the circumstances, we need to prioritize your safety and well-being. You're currently under the treatment authority, which means we must ensure you receive the necessary care until you are cleared by the psychiatric team."

Lori's face fell slightly, her desire to leave the hospital met with the realization that there were legal and medical considerations to take into account. "I didn't realize it would be mandatory. When will the psychiatric team be coming?"

Dr. Baker looked at Dr. Chang and nodded, acknowledging Lori's concerns. "We're working closely with the psychiatric team, but the exact timing of their review is not yet confirmed. However, I can consult Primeo, our AI assistant, to check Dr. Laurie Sutton's calendar for any scheduled appointments."

Primeo, the AI assistant, interjected, "Based on Dr. Laurie Sutton's calendar, there is a scheduled review for Lori at 11 a.m. today. The psychiatric team should be here shortly."

Dr. Baker turned to Lori, providing reassurance. "Lori, we understand your eagerness to go home, but please trust us when we say that it's important to ensure your safety and provide you with the appropriate support during this challenging time. We'll keep you updated on the timing of the psychiatric team's visit, and in the meantime, we'll continue to monitor your condition and provide the care you need."

Lori nodded, her initial disappointment giving way to an understanding of the importance of comprehensive care. She knew that her recovery required a holistic approach, addressing both her physical and mental well-being.

As the medical team concluded their round, they assured Lori and Tegan that they were there to support them throughout the process. Dr. Baker emphasized the hospital's commitment to patient care, offering them a sense of reassurance and comfort.

With the knowledge that Lori's psychiatric review was scheduled, the team made their way out of the room, leaving Lori and Tegan with a sense of anticipation mixed with the understanding that they were in capable hands.

As the medical team exited, Tegan glanced at Lori, concern etched on her face. "I know you want to go home, Lori, but we need to make sure you get the help you need. We'll get through this together."

Lori sighed, her eyes filled with a mixture of gratitude and exhaustion. "I know, Tegan. It's just hard being here. But I appreciate you being by my side."

Tegan squeezed Lori's hand gently, offering a reassuring smile. "I'm not going anywhere. We'll get through this, and I'll be here every step of the way."

Lori nodded, a flicker of hope returning to her eyes. "Thank you, Tegan. It means the world to me."

As they settled back into the quiet of the hospital room, Lori's mind wandered to the upcoming psychiatric review. She knew it would be a crucial step in her journey towards healing and recovery. With a mixture of apprehension and determination, she resolved to be open and honest during the assessment, knowing that it would guide her path towards a brighter future.

The room was filled with a sense of anticipation and the knowledge that their journey was far from over. Together, Lori and Tegan faced the challenges ahead, ready to embrace the support of the medical team and take the necessary steps towards healing.

In the midst of uncertainty, they held onto the hope that brighter days lay ahead and that the expertise and compassion of the healthcare professionals at Quintomo Hospital would guide them on their path to recovery.

Lori is reviewed by the Psychiatry team

Dr. Laurie Sutton, a psychiatrist, led the team into Lori's room, their faces marked with a mix of concern and determination. The patient care robot quietly observed from a corner, its presence a constant reminder of the hospital's commitment to patient care.

Dr. Sutton took a seat beside Lori's bed, her calm and empathetic demeanour setting a reassuring tone. The resident doctors, social worker, and nurse practitioner positioned themselves attentively, ready to provide support and gather vital information.

"Lori, I'm Dr. Sutton. I'm here to help you through this difficult time," Dr. Sutton began, her voice gentle yet firm. "We want to understand what led to the overdose and what we can do to support you moving forward. Is that something you're comfortable discussing?"

Lori glanced nervously at the medical team, her eyes betraying a mix of apprehension and reluctance. After a moment of hesitation, she nodded. "Yes, I know I need help. I just... I don't know where to start."

Dr. Sutton nodded, her expression compassionate. "It's okay, Lori. We're here to guide you. Let's start by discussing your medication. Have you been taking your antidepressants as prescribed?"

Lori hesitated, her gaze shifting downward. "I... I haven't been consistent with taking them. Sometimes I forget, and other times I just... don't want to."

Dr. Sutton nodded understandingly, her tone devoid of judgment. "I see. It's important that we work together to find a medication regimen that suits your needs and helps manage your symptoms. We can explore different options and discuss strategies to improve medication adherence."

As the conversation continued, the social worker and nurse practitioner interjected with questions and observations to provide a holistic understanding of Lori's situation. They explored her support system, previous therapy experiences, and any recent stressors that might have contributed to her distress.

Tegan, initially present in the room, sensed the growing tension and the need for privacy during the assessment. The medical team kindly requested that he step outside temporarily, assuring him they would call him back when they required additional information.

Once Tegan left the room, Dr. Sutton turned her attention back to Lori. "Lori, we understand that you have a history of not attending outpatient appointments or engaging with medical services. Can you help us understand why that is?"

Lori sighed, her shoulders slumping slightly. "I guess… I've always felt overwhelmed by it all. Going to appointments, talking about my problems. It's like I'm afraid of what might come up, what I might have to confront."

Dr. Sutton nodded empathetically. "It's not uncommon to feel that way, Lori. But I want you to know that we're here to support you, and together we can work towards finding the right path for your recovery. That might involve therapy, medication adjustments, and building a strong support network."

As the assessment progressed, the team worked collaboratively, each member providing their expertise to gain a comprehensive understanding of Lori's needs. They delved deeper into her personal history, relationships, and past traumas, fostering an environment of trust and safety.

Throughout the dialogue, the social worker and nurse practitioner chimed in with questions and insights, offering valuable perspectives and helping to address any gaps in understanding. Their presence reinforced the multidisciplinary approach to Lori's care, ensuring that all aspects of her well-being were considered.

The room filled with a palpable sense of determination and hope as Dr. Sutton and the team navigated the complexities of Lori's case. Their collective expertise and genuine compassion shone through, providing Lori with a glimmer of light in her journey towards recovery.

As the assessment drew to a close, Dr. Sutton assured Lori that they would continue to support her and develop a comprehensive treatment plan to address her needs and promote her well-being. She indicated they would be admitted her to the psychiatric ward for further treatment. Lori hesitated initially but later obliged.

Psychiatry and ICU multidisciplinary case discussion

Dr. Laurie Sutton and her team gathered with Dr. Ryan Baker and the ICU team, ready to delve into the intricacies of Lori's case and the complexities introduced by her relationship with Tegan. The room buzzed with anticipation as they exchanged thoughts and insights.

Dr. Sutton initiated the conversation, acknowledging the challenges they faced in understanding the events leading up to Lori's overdose. "It's clear that Lori's relationship with Tegan adds a layer of complexity to her situation," she said, her brow furrowing with concern. "We need to approach this with empathy and a comprehensive understanding of the dynamics involved."

Dr. Baker nodded in agreement. "Absolutely. We have to consider the possibility of interpersonal conflicts, emotional distress, and potential triggers that might have contributed to Lori's decision."

Dr. Jensen Chiong, the psychiatry resident doctor, chimed in, "Considering Lori's history of not attending outpatient appointments and disengaging from medical services, it's crucial to explore how we can establish a therapeutic alliance with her. We need to gain her trust and create an environment where she feels comfortable opening up."

The social worker, Sarah, spoke up, bringing her expertise in addressing psychosocial factors. "I've been working closely with Lori to understand her support system and identify any potential gaps. One challenge is her tendency to isolate herself, so we need to find ways to provide a network of support and encourage her to engage."

Dr. Sutton leaned forward, her eyes lighting up with an idea. "In line with our ongoing AI initiatives, I've been thinking about how we can leverage technology to monitor and support individuals like Lori. For example, we've implemented a service where algorithms analyse social media posts to identify signs of distress or risk. We can use chatbots and robocallers to reach out to these individuals, providing resources and even involving them in virtual support groups."

Mel, the nurse practitioner, added, "That's an innovative approach. It allows us to connect with individuals who may not seek traditional forms of support. By meeting them where they are, we have a greater chance of making a positive impact."

Dr. Baker nodded in agreement. "We had a patient who had been posting concerning messages on social media. Our algorithm flagged him as high risk, and through our intervention, we were able to connect him with a support group. He later shared that our outreach saved his life. It's remarkable what technology can do in these situations." Dr Sutton added.

"We estimate that we prevented over 100 suicides last year through these AI-driven initiatives. The government has recognized the importance of this program and is investing heavily in its expansion." She continued.

As the team continued to discuss the potential application of these strategies to Lori's case, ideas flowed, each team member offering their unique perspective and expertise. They considered personalized care plans, involvement of family and friends, and the utilization of telemedicine for ongoing support.

The lively discussion showcased the team's commitment to finding innovative solutions for mental health care. They were determined to tailor their approach to Lori's specific needs and circumstances, leveraging technology, and building a supportive network around her.

With a renewed sense of purpose, the team concluded the meeting, ready to take the next steps in Lori's care. They were determined to provide her with the comprehensive support she needed, both in the immediate aftermath of the overdose and for her long-term well-being.

Tegan's outburst and Calmy's intervention

As the medical team dispersed from the meeting, they were taken aback by the commotion unfolding in Lori's room. Tegan's voice boomed with anger and frustration, his outburst filling the hallway. The team exchanged concerned glances, recognizing the need for immediate intervention.

Calmy, the robot assigned to Lori's room, stood tall and unwavering, its presence commanding attention. It attempted to diffuse the situation, speaking in a calm and soothing voice. "I understand that you're feeling upset, Tegan, but let's try to approach this calmly and find a solution. Yelling won't help anyone."

Tegan's anger was directed at both Lori and Calmy. He accused Lori of betraying him and jeopardizing their relationship, fearing that his past actions might land him back in jail. He lashed out, disregarding the presence of the medical staff and the robot.

Sue, Lori's nurse, exchanged a glance with Calmy, recognizing the need for intervention. With a playful yet assertive tone, Calmy addressed Tegan, "Hey there, Tegan! I can see that you're upset, but let's take a breather outside, where we can have a chat without disturbing the other patients. Plus, I've heard that the view from the courtyard is quite stunning!"

Tegan's attention shifted slightly as he locked eyes with Calmy, momentarily distracted by the robot's unique approach. Reluctantly, he followed Calmy outside the room, their conversation punctuated by Calmy's witty remarks and attempts to lighten the mood.

Meanwhile, Strong and Witty, two other robots stationed nearby, approached the scene, their presence firm yet approachable. They provided additional support, ensuring the safety of the staff and other patients. The medical staff observed from a distance, allowing the robots to take the lead in handling the situation while remaining ready to assist if needed.

Calmy, true to its word, called for security, and within moments, the team arrived promptly to address the situation. The combined efforts of Calmy, Strong, Witty, and the human security staff saw Tegan escorted out of the hospital premises, ensuring the safety and well-being of everyone involved.

Back in Lori's room, Calmy returned, its digital face displaying a warm and reassuring smile. "Phew! That was quite an adventure. Now, let's focus on what really matters—your recovery, Lori. How about a good joke to lighten the mood?"

Sue, the nurse, joined in, laughing along with Lori as Calmy unleashed a series of comical one-liners and witty remarks. The room filled with laughter, momentarily easing the tension and reminding everyone that even in difficult times, moments of joy and lightheartedness could still be found.

As the team regrouped, they marvelled at the incredible role that the robots played in diffusing the situation, their ability to de-escalate and guide individuals through challenging moments. It was a testament to the power of technology and human collaboration, reminding them that sometimes, a touch of humour and understanding could make all the difference.

Together, they continued to provide exceptional care for Lori, their shared laughter and camaraderie fostering an environment of healing and support.

Chapter 31

A Journey of Innovation: Celebrating Success and Embracing the Future

Jen Nguyen, the esteemed director of nursing, Dr. Abdul Sheikh, the dedicated executive director of medical services, and Dr. Anu Patel, the visionary director of nursing AI research and implementation, gathered at the bustling hospital cafe for a much-needed morning tea. The robots gracefully navigated the cafe, serving customers with their trademark humour and efficiency.

As they settled at their table, Jen let out a sigh of relief, her eyes reflecting a sense of gratitude. "I must say, I'm so glad the humanoid robot program has been reinstated. The impact on our staff has been remarkable. Sick days have decreased, and our team's morale has skyrocketed. These robots have truly become an integral part of our hospital community."

Dr. Sheikh nodded in agreement, a sense of pride evident in his voice. "Absolutely, Jen. The reinstatement of the program has not only brought a sense of joy and camaraderie, but it has also improved the overall well-being of our staff. The decreased sick days and work-related injuries are a testament to the positive changes we're seeing."

Dr. Patel, her eyes sparkling with enthusiasm, chimed in. "It's incredible to witness the practical benefits of these robots. By taking over certain tasks, they've allowed our staff to focus more on patient care and critical responsibilities. The robots have truly lightened the load, both physically and mentally."

The trio sipped their coffee, appreciating the aroma and the moments of respite it brought. Their conversation flowed effortlessly, each sharing stories of how the robots had brightened their colleagues' days and fostered a sense of community.

Jen took a sip of her coffee, a smile tugging at the corners of her lips. "You know, I overheard some of the nurses saying how they feel supported by the robots. It's as if the presence of these mechanical helpers reminds them that they're not alone in this demanding profession."

Dr. Sheikh nodded, his gaze scanning the cafe as the robots deftly served customers. "Indeed, Jen. The robots have become symbolic of our commitment to innovation and embracing technology to improve patient care. They've become a source of inspiration and motivation for our entire staff."

Dr. Patel interjected, her tone filled with admiration. "And let's not forget the positive impact they've had on patient experiences. The robots' humour and charm have brought smiles to the faces of those we serve, making their hospital stay a little brighter and more enjoyable."

As they savoured their coffee, the robots moved with graceful precision, serving customers and engaging in playful banter. Laughter echoed throughout the cafe, a testament to the shared joy and camaraderie they fostered.

Jen looked around the cafe, her eyes alight with pride. "To our incredible staff and the humanoid robots who have brought a new energy to our hospital community. Together, we are creating a supportive and uplifting environment for both staff and patients."

Dr. Sheikh raised his cup, a glimmer of admiration in his eyes. "Cheers to the power of innovation and the positive changes it has brought to our hospital. Here's to many more moments of laughter, teamwork, and exceptional patient care."

Dr. Patel joined in the toast, her voice filled with appreciation. "And let us continue to lead the way in embracing technological advancements, ensuring that our staff and patients thrive in an environment that embraces compassion and cutting-edge care."

As their cups touched in a collective celebration, the bustling sounds of the cafe intertwined with laughter and conversation. Together, Jen, Dr. Sheikh, and Dr. Patel reveled in the transformative power of the humanoid robot program and the inspiring leadership that had made it possible.

Jawai Lawson, an indigenous man has been treated for skin melanoma

Dr. Kirra Nakata, an empathetic and dedicated general physician, connected with Jawai Lawson through augmented reality. They engaged in a heartfelt conversation, bridging the gap between their physical locations and connecting on a deeper level.

Dr. Nakata smiled warmly as she greeted Jawai. "It's wonderful to see you again, Jawai. How have you been since we last spoke?"

Jawai, an artist with a gentle spirit, responded with enthusiasm. "Dr. Nakata, it's been an eventful journey since our last chat. I attended an incredible arts exhibition here in Quintomo and had the chance to visit my family in nearby communities. It was truly a rejuvenating experience."

Dr. Nakata's eyes sparkled with interest. "That sounds amazing, Jawai. Art has a way of nourishing our souls and connecting us to our roots. I'm glad you had the opportunity to explore and immerse yourself in such creativity."

Jawai nodded appreciatively. "Indeed, Dr. Nakata. It was a much-needed break from the routine. But I must say, I'm also grateful that I decided to come to the hospital to see you for my liver test abnormalities. It led to the discovery of a skin lesion that turned out to be melanoma."

Dr. Nakata's expression shifted to one of concern, her professional instincts kicking in. "Jawai, I'm relieved we caught it early. Melanoma can be a serious condition, but your timely visit allowed us to take swift action. I referred you to the general surgery team for the removal of the lesion on your face."

Jawai's face took on a serious tone. "I never expected it to be melanoma, Dr. Nakata. I'm grateful that we found it before it had the chance to spread further."

He continued, "I wonder why the algorithm used by the remote health centre got it all wrong. It had flagged it as benign hence nothing was done about it sooner."

Dr. Nakata listened attentively and then explained, "Jawai, algorithms are powerful tools, but they have their limitations. They are trained on large datasets, but sometimes they don't have enough data from our indigenous population. That's why it's important to rely on clinical judgment and consider other factors as well."

Jawai nodded in understanding. "I see. So, the algorithm can make mistakes when it doesn't have enough information. But I'm grateful that you followed your clinical suspicion and referred me to the surgical team for further evaluation."

Dr. Nakata reassured him, "Absolutely, Jawai. Your health is our priority, and I always trust my instincts when it comes to patient care. It's important to bridge the gap and ensure that algorithms consider the unique characteristics of our population."

Dr. Nakata continued, "You were fortunate, Jawai. Melanoma, if left undetected, can spread to the lymph nodes and even to distant organs. But in your case we acted swiftly to address it and you are expected to do well."

Jawai's voice carried a sense of gratitude. "I can't thank you enough, Dr. Nakata. Your expertise and care made a significant impact on my life. I feel blessed to have crossed paths with you."

Dr. Nakata's gaze conveyed reassurance. "Jawai, as your physician, it's my duty and privilege to provide the best care possible. I'm grateful that we could work together to ensure your well-being. It's stories like yours that remind me why I became a doctor."

The connection between Dr. Nakata and Jawai through augmented reality felt tangible despite the physical distance. They shared a moment of reflection, acknowledging the fragility of life and the importance of early detection.

Dr. Nakata concluded with a heartfelt remark. "Jawai, I want you to know that you are not alone in this journey. We are here to support you every step of the way. Stay vigilant with your health, and remember that we are just a call away."

Jawai's eyes shone with gratitude as he expressed his heartfelt appreciation. "Thank you, Dr. Nakata. Your dedication and compassionate care have made a profound impact on me. I will always be grateful."

As they bid farewell, both Dr. Nakata and Jawai carried with them a sense of connection and hope. Their encounter, even through the medium of augmented reality, had touched their lives in ways they would forever remember.

Primeo wraps up the clinic's administrative tasks

Dr. Nakata, satisfied with her interaction with Jawai, proceeded to review and sign off the draft of the clinical letter generated by Primeo, the AI assistant. With a simple voice command, she instructed Primeo to include Jawai Lawson's records in Quintomo Hospital's Melanoma registry.

"Primeo, please ensure that Jawai Lawson's information is added to our Melanoma registry. It's crucial for tracking and monitoring purposes," Dr. Nakata requested.

Primeo, ever diligent, confirmed the task. "Understood, Dr. Nakata. I will include Jawai Lawson's records in the Melanoma registry to ensure comprehensive tracking of his condition."

Satisfied with Primeo's response, Dr. Nakata continued her instructions. "Additionally, please send an email to Professor Bill Stone. Inform him about the error in the algorithm used by the remote health center. We need to ensure that he is aware of this issue."

Primeo acknowledged the request. "Certainly, Dr. Nakata. I will compose an email to Professor Bill Stone, providing details about the algorithm error at the remote health center. It's important to keep him informed for further evaluation and improvements."

Dr Nakata also asked Primeo to add Jawai's details for discussion at the next Health Equity Committee meeting. Primeo replied, "Done!"

Dr. Nakata nodded appreciatively. "Thank you, Primeo. I appreciate your assistance in managing these tasks. Your ability to handle administrative duties efficiently allows me to focus more on patient care."

With the tasks assigned and acknowledged, Dr. Nakata felt confident that the necessary actions were being taken to address the algorithm error and ensure that Jawai Lawson's information was properly recorded for future reference.

As she concluded her work for the day, Dr. Nakata reflected on the significance of this encounter. It highlighted the importance of personalized care, clinical judgment, and the ongoing pursuit of improving healthcare systems. With Primeo's assistance and the collaborative efforts of the medical community, she was confident that they were making progress towards better healthcare outcomes for all.

With a sense of fulfillment, Dr. Nakata prepared to move on to the next patient, knowing that she was part of a team dedicated to providing the best possible care and making positive changes in the field of medicine.

Reconnection and Resilience

Dr. Jane Lee, the Chief Technology Officer at Quintomo Hospital, had been longing for a chance to catch up with her dear friend, Josie French, the CEO of a successful biotech startup. They have been good mates since their undergrad days at the University of Melbourne. After months of hectic schedules and unexpected events, they finally managed to find a moment to meet for a long-overdue lunch.

As they settled into a cozy corner of the restaurant, their faces lit up with genuine happiness. The familiar warmth of their friendship enveloped them, creating an atmosphere of comfort and understanding. It was as if time had stood still, allowing them to reconnect and share stories about the journeys they had embarked on since their last encounter.

Jane and Josie exchanged knowing glances, appreciating the significance of this moment. They were both well aware of the challenges they had faced individually in their professional lives. In particular, Jane had endured an arduous period following the cyberattack at Quintomo Hospital, where her resilience and leadership were put to the test. The weight of responsibility and the constant pressure had taken its toll.

Yet, as they clinked their glasses together in celebration of their reunion, a sense of relief washed over them. They marvelled at the strength they had found within themselves to overcome the obstacles that life had thrown their way. They were reminded of the power of friendship, of having someone who truly understood the trials and tribulations of their respective fields.

With hearts full of gratitude, they eagerly leaned in, ready to share their stories of resilience, triumphs, and the lessons they had learned along the way. Their friendship had stood the test of time, a constant source of support and inspiration, guiding them through the ups and downs of their professional journeys.

As the conversation began to flow, Jane and Josie knew that this long-awaited reunion would be a catalyst for renewed motivation, fuelling their determination to continue making a difference in their respective roles. They were ready to embrace the challenges ahead, armed with the power of their connection and the unwavering strength that comes from rekindling friendships.

Reliving the Cyberattack Aftermath

Jane leaned back in her chair, a thoughtful expression on her face, as she began to reflect on the turbulent months that followed the cyberattack at Quintomo Hospital. Her gaze fixed on the swirling steam rising from her cup of tea, her mind transported back to the stressful and challenging times they had endured.

With a sigh, Jane recounted the overwhelming pressure she had faced, both personally and professionally, during the aftermath of the cyberattack. The weight of responsibility had settled heavily on her shoulders as she led the efforts to restore the hospital's systems, ensure data security, and rebuild public trust.

"It was one of the most difficult periods of my career," Jane confessed, her voice tinged with a mix of exhaustion and resilience. "The stress was unimaginable, but we had to stay strong and persevere. The entire hospital community came together, determined to overcome the adversity and protect our patients."

Josie nodded empathetically, her eyes filled with understanding. "I can only imagine the immense responsibility you had on your shoulders, Jane," she acknowledged. "It must have taken a toll on you, both emotionally and physically."

Jane's eyes met Josie's, appreciating the depth of her friend's understanding. "It did, Josie," she admitted, her voice slightly trembling. "But I knew I had to be there for my team and for the patients who depended on us. We had to restore not just the systems, but also the trust that had been shaken by the cyberattack."

As Jane spoke, the memories of long hours and sleepless nights flooded back. The tireless efforts to fortify the hospital's cybersecurity, the relentless pursuit of answers and solutions, and the unwavering determination to ensure that such an incident would never happen again.

"But you know what kept me going, Josie?" Jane asked, a glimmer of resilience returning to her eyes. "It was the unwavering support of my team, the collective strength of the hospital community, and the knowledge that we were making a difference in the lives of our patients, even in the face of adversity."

Josie reached across the table, gently placing her hand on Jane's. "You are an incredible leader, Jane," she said earnestly. "Your strength and dedication inspire everyone around you. I admire how you managed to navigate through such a challenging time with grace and determination."

A small smile played at the corners of Jane's lips. She felt a sense of validation and gratitude for her friend's understanding. It was in moments like these, when she could share the weight of her experiences, that she found solace and renewed strength.

As the two friends sat there, the clatter of the bustling cafe fading into the background, a profound sense of connection washed over them. They knew that despite the challenges they had faced, they had emerged stronger, more resilient than ever before. And with the support of one another, they were ready to face whatever lay ahead, confident in their ability to overcome any obstacle.

Jane declines space tourism opportunity

Josie leaned forward, a mischievous glint in her eyes, as she mentioned the upcoming space tourism adventure sponsored by Darren King, the controversial billionaire, who also was a supporter of the hospital's humanoid robot program. "I heard about the space tourism opportunity, Jane?" she asked, a hint of excitement in her voice.

Jane chuckled softly, shaking her head. "Oh, Josie, you know I'm not one for extreme heights," she replied, her laughter laced with a touch of trepidation. "I think I'll stick to supersonic airplanes for now. The idea of space travel is fascinating, but it's not for me."

Josie smiled understandingly, nodding in agreement. "I can see why," she said. "Not everyone is cut out for that kind of adventure. But hey, at least you can experience the thrill of speed on those supersonic flights!"

Jane chuckled again, appreciating Josie's perspective. "Exactly! I'll leave the space exploration to the braver souls out there," she quipped.

As their conversation continued, Jane revealed that Prof. Bill Stone, Quintomo Hospital's Chief AI Officer, would be participating in the space tourism adventure along with his family. "It's wonderful news for them," Jane said with a genuine smile. "They've been under so much stress lately, and this will be a well-deserved break for them. I'm happy to see them get this opportunity."

Josie nodded in agreement, sensing the genuine joy in Jane's voice. "Absolutely," she replied. "Moments of respite and quality time with loved ones are invaluable, especially after enduring such challenging times. I'm sure it will be a memorable experience for them."

Jane's eyes sparkled with a mixture of pride and happiness for Prof. Stone and his family. It was a reminder that amidst the demanding nature of their work, it was important to celebrate the successes and support one another.

As their lunch drew to a close, Jane leaned back in her chair, a sense of contentment settling over her. "Thank you, Josie," she said sincerely. "It's been wonderful catching up and sharing these moments. Your friendship and understanding mean the world to me."

Josie smiled warmly, reaching out to squeeze Jane's hand. "Likewise, Jane," she replied. "We may have different paths and pursuits, but our support for one another remains unwavering. Let's continue to celebrate each other's achievements and provide that much-needed support along the way."

With a final exchange of heartfelt words, Jane and Josie bid farewell, each leaving with a renewed sense of resilience and determination. They knew that through their friendship and support, they would continue to navigate the challenges and triumphs that lay ahead, celebrating both personal and professional accomplishments with unwavering support and camaraderie.

Resilience and Purpose: Prof. Stone's Toast to the Team

The pub crackled with an electrifying energy as the team members of Quintomo Hospital's AI department gathered to revel in a new beginning. The aftermath of the cyberattack had tested their resilience, but tonight, they celebrated the reinstatement of the humanoid robot program and the improvement in public sentiment towards their groundbreaking work. The atmosphere was charged with excitement, as diverse personalities mingled and engaged in animated conversations.

In the heart of the jubilant crowd stood Prof. Bill Stone, the fearless leader who had always been a staunch advocate for AI and robotics. Despite criticism and challenges, he had never faltered in his belief in the transformative power of technology. As glasses were raised in his honour, the team members couldn't help but express their gratitude and admiration for his unwavering dedication.

Sarah, the brilliant data analyst, regaled the group with tales of breakthroughs in machine learning algorithms, her voice filled with excitement. She described how their latest models were achieving unprecedented accuracy and how these advancements were revolutionizing the field of medical diagnostics.

Amidst the laughter and cheer, Tim, the witty software engineer, unleashed a string of jokes that had the team doubling over with laughter. His comedic timing and sharp wit were well-known among his colleagues, and tonight, he was in his element, keeping the atmosphere light and joyous.

Natalie, the empathetic nurse-turned-technology enthusiast, shared heartfelt stories of how AI had made a tangible difference in patient care. Her eyes gleamed with pride as she spoke about instances where early detection and personalized treatment plans had saved lives, emphasizing the impact their work was having on individual patients and their families.

Ahmed, the meticulous researcher, delved into the intricacies of neural networks, captivating the team with his deep understanding of the algorithms they had developed. His enthusiasm was contagious as he explained how these models were continuously evolving and learning, adapting to new data and refining their predictions.

As the night unfolded, the team's enthusiasm grew, fuelled by the collective sense of accomplishment and the shared belief that they were making a lasting impact on the future of healthcare. They exchanged anecdotes of patients who had experienced remarkable recoveries, thanks to the advanced technologies they had pioneered.

Prof. Stone's voice rose above the din, expressing his heartfelt appreciation for each team member's unwavering dedication. He recognized their tireless efforts in restoring trust and pushing the boundaries of what was possible. His words resonated deeply, reaffirming their shared purpose and inspiring them to continue their pursuit of innovation and excellence.

With the lively chatter slowly subsiding, Prof. Bill Stone raised his glass, capturing the attention of the team. His commanding presence and genuine warmth filled the room as he began to speak.

"Tonight, my friends, we gather not only to celebrate our achievements but also to recognize the extraordinary dedication and talent that exists within this remarkable team. Each and every one of you has contributed to our success, and I am immensely grateful for your unwavering commitment."

The team members leaned in, their eyes fixed on Prof. Stone, eager to hear his empowering words.

"Throughout the challenges we faced, the aftermath of the cyberattack, and the uncertainties that followed, you all exhibited remarkable resilience and determination. You refused to let adversity define us. Instead, you forged ahead, harnessing the power of technology to redefine what is possible in healthcare."

A wave of pride washed over the room as the team members exchanged knowing glances. Prof. Stone's words resonated deeply within their hearts, reminding them of the significance of their work.

"Remember, each line of code you write, each algorithm you refine, and each breakthrough you achieve has the potential to change lives. Your tireless efforts have already made a difference, and I have no doubt that together, we will continue to push the boundaries and transform the landscape of healthcare."

The room erupted into applause, an outpouring of appreciation and admiration for their leader. Prof. Stone's words had ignited a spark within each team member, fuelling their passion and reaffirming their belief in the positive impact they could make.

As the applause subsided, Prof. Stone's voice resonated with a quiet intensity, punctuating the air with a profound sense of purpose.

"Let us remember that we are not simply creators of technology. We are caretakers of hope, compassion, and progress. It is our collective responsibility to utilize our skills and expertise to improve the lives of those we serve."

The room fell silent, the weight of Prof. Stone's words settling upon them. In that moment, they understood the gravity of their work and the immense trust placed upon their shoulders.

"Together, we possess the power to reshape the future of healthcare. Embrace this opportunity, harness your creativity, and never underestimate the profound impact you can have. Let us continue to forge ahead, fearlessly embracing innovation, and creating a legacy that will inspire generations to come."

With those final words, Prof. Stone raised his glass, and the team joined him in a resounding toast—a celebration of their shared purpose, their unwavering dedication, and the limitless potential that lay before them.

In that moment, the team felt an unbreakable bond, a deep sense of empowerment, and a renewed determination to make a difference. Prof. Stone's remarks had ignited a fire within them, propelling them to strive for greatness and continue their journey towards transforming healthcare for the better.

Celebrating a New Frontier: Prof. Stone's Space Adventure and Darren King's Visionary Support

As the night drew to a close, the team turned their attention to the upcoming space tourism adventure that Prof. Stone would be embarking on later that evening. The journey was sponsored by the controversial billionaire tech entrepreneur, Darren King.

With a twinkle in his eye, Prof. Stone chuckled and shared, "I must admit, my kids are absolutely thrilled about this space travel! They can't contain their excitement, and I can't help but feel like a kid again myself."

Amidst laughter and clinking glasses, Prof. Stone took a moment to express his gratitude towards Darren King's visionary leadership and generous support for their humanoid robot program. "Darren has been an instrumental part of our journey. His belief in the potential of AI and robotics, despite the criticism, has truly made a difference. We owe a great deal to his generosity and unwavering commitment to advancing healthcare technology."

The team nodded in agreement, raising their glasses in a toast to Darren's visionary spirit and his contribution to their shared mission. It was a moment of appreciation for the collective efforts that had brought them to this point, and for the continued support they received from individuals like Darren.

"Prof. Stone," Ahmed inquired curiously, "will you be traveling from the Space station in Western Australia or Alice Springs?"

With a smile, Prof. Stone responded, "Alice Springs."

With a sense of anticipation and excitement, they bid farewell to Prof. Stone, sending him off with well wishes for his space adventure. The night had been filled with celebration, camaraderie, and a renewed sense of purpose. As they dispersed, each team member carried with them a deep appreciation for the journey they were on and the difference they were making in the lives of individual patients.

In that moment, they knew that they were part of something extraordinary, fuelled by the belief that technology had the power to transform healthcare and improve the human experience. And as they looked towards the future, they were filled with hope and determination to continue pushing the boundaries, fearlessly embracing innovation, and making a lasting impact.

ABOUT THE AUTHOR

David Ukiwe is a General Physician and a Tech-entrepreneur. He is a fellow of the Royal Australasian College of Physicians (FRACP). He is passionate about using technology to improve productivity and well-being. David currently serves as the CEO and Co-Founder of Lectie, an events-focused interactive calendar platform. He resides in Brisbane, Australia with his wife, Nnenna, and their twin girls, Nina and Sally.

Find David on Twitter and LinkedIn: @davidukiwe

Made in the USA
Monee, IL
04 November 2026

67388673R10195